# VENICE REDISCOVERED

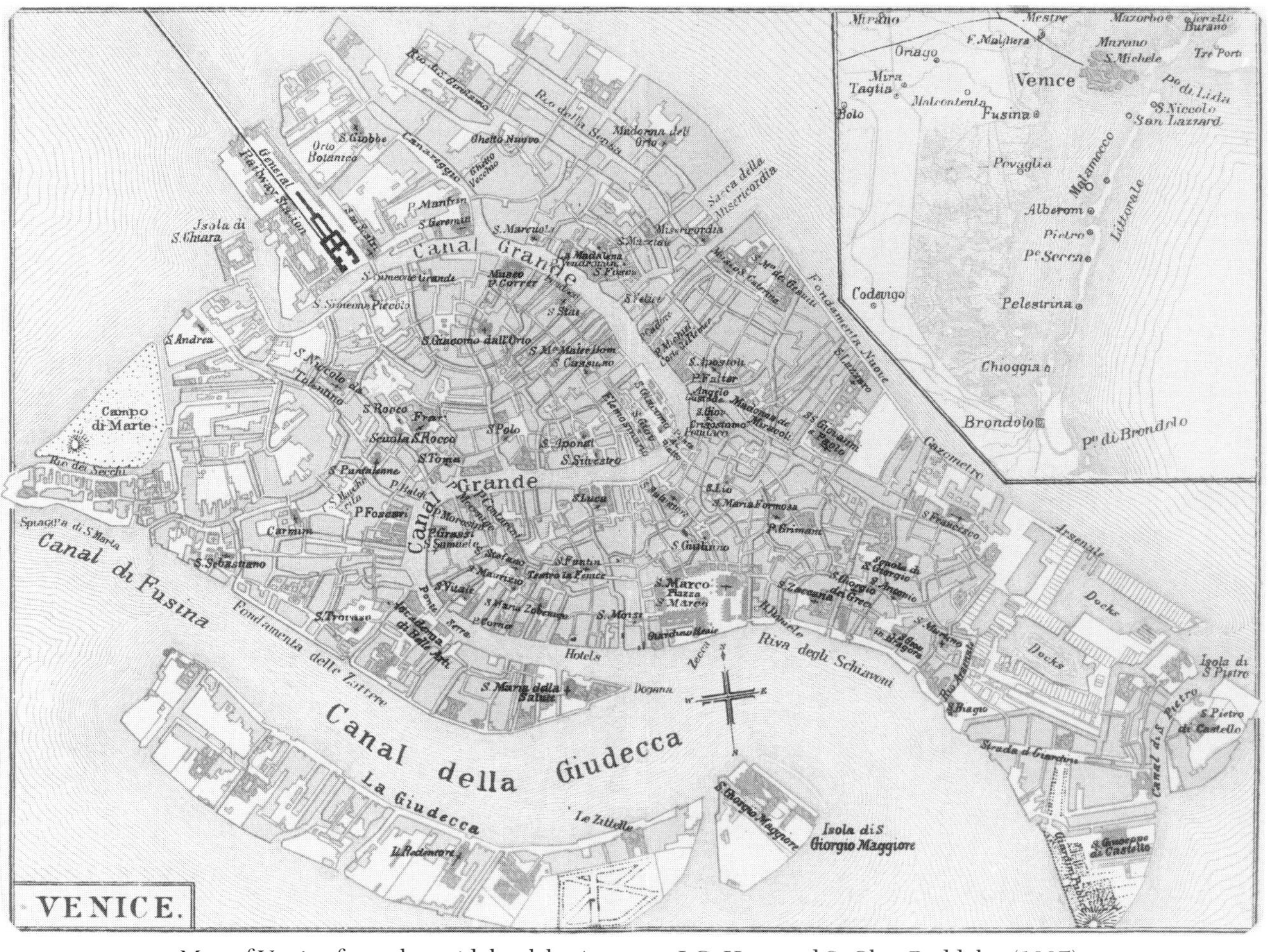


Map of Venice from the guidebook by Augustus J.C. Hare and St Clair Baddeley (1907).

# VENICE REDISCOVERED

John Pemble

Man is in love and loves what vanishes,
What more is there to say?

W. B. Yeats

CLARENDON PRESS · OXFORD

*Oxford University Press, Walton Street, Oxford OX2 6DP*
*Oxford New York*
*Athens Auckland Bangkok Bombay*
*Calcutta Cape Town Dar es Salaam Delhi*
*Florence Hong Kong Istanbul Karachi*
*Kuala Lumpur Madras Madrid Melbourne*
*Mexico City Nairobi Paris Singapore*
*Taipei Tokyo Toronto*
*and associated companies in*
*Berlin Ibadan*

*Oxford is a trade mark of Oxford University Press*

*Published in the United States*
*by Oxford University Press Inc., New York*



*British Library Cataloguing in Publication Data*
*Data available*

*Library of Congress Cataloging in Publication Data*
*Pemble, John.*
*Venice rediscovered / John Pemble.*
*p. cm.*
*Includes bibliographical references and index.*
*1. Venice (Italy)—Civilization. I. Title.*
*DG675.6.P36 1995*
*945'.31—dc20 94–17844*
*ISBN 0–19–820501–5*

3 5 7 9 10 8 6 4 2

*Printed in Great Britain on acid-free paper by*
*Bookcraft Ltd., Midsomer Norton, Avon*

For John and Nicholas

# ACKNOWLEDGEMENTS

MY first debt of gratitude is to the Leverhulme Trust, for a research fellowship that financed study leave, and for the support and encouragement that made the project seem worthwhile. I also acknowledge with thanks assistance towards the cost of research in Venice from the Gladys Krieble Delmas Foundation of New York, and a grant for purposes of research and travel in Britain from the Arts Faculty Research Fund of the University of Bristol.

Help and advice were freely given by Mrs Betty Coley of the Armstrong Browning Library, Baylor University, Texas; by Messrs Nick Lee and Michael Richardson in the Special Collections, University of Bristol; and by the staffs of the Biblioteca Marciana, Bristol Central Library, Bristol University Library, the Bodleian Library, the British Library, the Brotherton Library, Cambridge University Library, the Dorset Records Office, the House of Lords Record Office, the National Library of Scotland, Nottingham University Library, Leeds District Archives, the Scottish Record Office, and Sheffield Central Library. Special thanks are due to Mrs Jean Bradford and the staff of the Inter-Library Loans unit at Bristol University Library. Permission to use copyright material was generously given by Mr Chris Furse OBE and Mr D. H. Brown of Pallamallowa, New South Wales. Expert advice and valuable suggestions came from Dr Tony Antonovics, Mr Nick Azis, Dr Alan Bell, Dr Peter Coates, Dr Basil Cottle, Professor Sir Geoffrey Elton, Dr Rowena Fowler, the late Dr John Leslie, Dr Graziella Mazza, Ms Alison Richards, Professor Jack Simmons, Professor Tony Tanner, Dr Hugh Tulloch, and Professor John Vincent. Without their help my difficulties would have been much greater and my errors much more numerous. Credit for any merit the work may have must be shared by the publisher's readers, who were forbearing and constructive in their response to the crudities of my original text. Acknowledgement is also due, and gratefully made, to Mr Ross Lamont, of the University of New England, New South Wales, for taking so much trouble to seek out and put me in touch with the family of Horatio Brown. Mrs Mary Tosh coped heroically with the tedious job of transforming a primitive typescript into professional word-processed copy; and the expert editorial staff of Oxford University Press spared no effort in working to turn the copy into a volume. No author could receive better treatment than I have received from OUP's history editor, Dr Tony Morris. Mine might have been the only progeny demanding his skills as accoucheur.

Dame Janet Vaughan figures in a special way among these claimants to my thanks. She shared with me some of the memories of her ninety years, including those of Horatio Brown, Edmund Gosse, and her grandmother, mother, and aunts—the widow and daughters of John Addington Symonds. To talk with her was to cross the threshold of an extraordinary past, full of Bloomsbury friendships and eminent Victorian survivals. I valued her kindness and encouragement, and had hoped to present her with a copy of this book. Sadly she died shortly before it was finished, and it remains without the approval that she was uniquely qualified to give.

J.P.

*Clifton*
*March 1994*

# CONTENTS

# LIST OF PLATES

*Between pages 84 and 85*

# INTRODUCTION

JULES AND EDMOND DE GONCOURT found it remarkable that Rousseau had spent more than a year in Venice in the 1740s and remained insensitive to its enchantment. In spite of his penchant for description, he had not seen the exquisite city that they and their contemporaries saw. They deduced therefore that the modern age had in some way clarified human vision. 'Le XIXe siècle a opéré l'humanité de la cataracte.'[1]

Their deduction was correct. The change that divided their generation from that of Rousseau was a change in perception rather than in the thing itself. Venice was not radically transformed in the nineteenth century, in the way that Rome was, and Paris, and London. If Canaletto, who died in 1768, had returned a hundred years later—when the Goncourts made their comment—he would have noticed many alterations, but he would not have felt a stranger in the city whose image he had made familiar abroad. During the next hundred years change was to be even less apparent. Yet appreciation of its beauty then became so acute, both in Europe and North America, that a passionate battle was fought and won to fabricate for Venice the illusion of immortality. When buildings crumbled or collapsed, they were resurrected by the thaumaturgic power of nostalgia. A rare example of conspicuous change was in the increasing dilapidation and shabbiness of Venice, especially in its outlying areas. The change in perception is registered by the fact that this evidence of deterioration made the city not less but more attractive. In the early 1800s Venice had been generally regarded as an odd and rather depressing wreck which could qualify as beautiful only when seen at a distance or by moonlight. By the end of the century the most fastidious sensibility was not only able but eager to contemplate the detail of its ruin. A transfiguring myth had developed, rooted in esoteric cults of art and literature; and as those cults became obsolete, the metabolism occurred that converts yesterday's highbrow conceit into today's middlebrow cliché. The myth lived on, providing a language and an iconography for advertising, journalism, and mass entertainment.

The cult of Decadence, whose origins were French but whose appeal spread throughout Europe, represented the intellectual revolt against Rousseau in his best-known role, as the apostle of nature. Gautier and Baudelaire despised nature, which they saw as the source of ugliness and dissonance; and French writers of the next generation, deeply humiliated by the disasters of the Franco-Prussian War, discovered a creed for their times in a mixture of Schopenhauerean pessimism and Baudelairean celebration of the artificial and the occult.[2] They yearned with Baudelaire for 'anywhere out of the world'; and it was in the city, that abomination of Rousseau, that they discovered their antithetical universe. The city was a theatre of masks and *maquillage*; a temple of the abnormal and the perverse; a hospital of pathological process. And Venice was the quintessential city. There were no slums more slummy than the Venetian back canals, with their leprous buildings and odour of decay; while in the great Venetian palaces there was an unparalleled example of human contrivance at odds with nature. No further refinement of art was possible. 'I do not understand why anyone paints Venice,' wrote the English poet of Decadence, Arthur Symons, 'yet everyone who paints, paints Venice . . . To do so is to forget that it is itself a picture, a finished, conscious work of art. You cannot improve the picture as it is, you can add nothing, you need arrange nothing. Everything has been done . . .' To Symons the Piazza San Marco suggested nothing so much as stage-scenery: 'I seemed, after all, not to have left London, but to be still at the Alhambra, watching a marvellous ballet . . . The Doge's Palace looked exactly like beautifully painted canvas, as if it were stretched on frames, and ready to be shunted into the wings for a fresh "set" to come forward.'[3]

He saw a stage that was deserted—'the actors, the dancers, are gone'. But they were not absent for long. The twentieth century brought to the city the film-makers, with their retinues of actors in make-up and costume. In 1922 the Italian writer Ugo Ojetti watched the shooting of *I Due Foscari*, one of the first of countless films to use Venice as a location, and relived Symons's experience of the real becoming fake: 'Il peggio si è che a fissare per mezz'ora quella mascherata al sole, noi stessi perdiamo il senso della solida realtà, e non le mura del palazzo [ducale] ci fanno sembrare vere le comparse, ma le comparse ci fanno sembrare finto il palazzo.'*[4]

Existing in symbiotic connection with this thirst for artifice and simulacrum, provoking it and being provoked by it, was an obsession with truth and fact. The age of Decadence overlapped the age of natural science and scientific history. The work of Niebuhr, Ranke, Michelet, Taine, Froude, Arnold, Milman, Acton, and the other lights of 'le siècle de l'histoire'[5] now lies, as in a mass grave, in the undisturbed recesses of older academic libraries. Yet nineteenth-century historiography is not entirely

* The worst of it is, that by watching for half an hour this masquerade in the sun, we ourselves lose the sense of solid reality, and instead of the walls of the [Ducal] Palace making the actors seem real, it is the actors who make the palace seem fake.

dead, because the modern appreciation of Venice is in some measure its legacy. In Rousseau's time Venice was an independent republic whose past and present reputation precluded sentimental rhapsody. By the time of the Goncourts the city had become a phantom, a relic, whose political power was extinct and whose history had been rewritten. Tyranny had gone; and in its place was the pathos of merit traduced and majesty dethroned. The nineteenth-century historians brought about this change of perception by looking at Venetian history not from the outside, but from the inside. That is to say, they used as evidence not the reports of foreign observers and contemporary chroniclers, but the records of the Venetian state itself. Furthermore they judged the Venetian Republic by a different light. They judged it not by the standards of a notional universal morality, but by the standards of the period they were discussing. Eighteenth-century historians—Gibbon, Voltaire, Turgot, Condorcet, Hume—had used the comparative method in order to discover an invariable. They had looked, amidst the accidentals of time and place, for human nature—something they assumed to have been always and everywhere the same. 'Mankind are so much the same', wrote Hume, 'in all times and places, that history informs us of nothing new or strange in this particular. Its chief use is only to discover the constant and universal principles of human nature.'[6] To the historians of the following century the results of such inquiry seemed highly suspect. The older historians were accused of having found what they wanted to find—which was a universal man who was rational, benevolent, and happy. They had used history to validate Rousseau's assumption about the natural goodness of man and the corrupting influence of society. The new historiography rejected Hume in favour of the Italian philosopher Giambattista Vico. By adopting Vico's view that human nature was subject to psychological and moral development, it introduced relativity into historical judgement and made possible the rehabilitation of discredited institutions. The historian, said Henry Hart Milman, should be superior to that 'contemptuous wisdom . . . which refers everything to one standard'.[7] The scientific methods of research perfected in the German universities were therefore applied by historians who abjured absolute morality. Both ancient and modern history were looked at afresh, from viewpoints that were themselves within history.

The growing attractiveness of Venice, then, can be explained in terms of new ways of looking at the city and at the past. But it was a question of more than just the discarding of eighteenth-century prejudice. Venice in its posthumous years acquired a poignant topicality. It had qualities that answered to the most deeply felt appeal of the modern heart—the appeal for permanence and coherence in a fragmenting and chaotic universe. There is copious evidence of intellectual suffering in nineteenth-century literature. This was the age of the *mal du siècle*: a neurosis of deracination and dislocation, caused by traumatic severance from the past and compounded by the prevailing drift of thought. What Nietzsche called the Socratic spirit was at work, preaching that virtue is happiness and that knowledge is virtue. It was pulling the world apart, and then failing to put it together again. And Nietzsche, the tormented thinker who was at the same time both a hater of Socratism and one of its most brilliant practitioners, represents the divided psyche of his century. The intellectual landscape was a battleground between the principles of analysis and synthesis, and analysis won all the victories. Elaborately constructed systems, like those of Comte and Spencer, built on the wreckage of destroyed faith, fell victim in their turn to the prevailing blight of scepticism. 'Il n'a pas inventé grand chose, ce misérable siècle', wrote the novelist J. K. Huysmans in 1891. 'Il n'a rien édifié et tout détruit.'*[8] The English historian James Anthony Froude remembered his youth as an initiation into an era of doubt: 'All around us the intellectual lightships had broken from their moorings, and it was then a new and trying experience. The present generation . . . will never know what it was to find the lights all drifting, the compasses all awry, and nothing left to steer by except the stars.'[9] As Nietzsche pointed out, a longing for art was the outcome of the great metaphysical illusion of Socratism.[10] When science reached those outer limits of inquiry where logic collapsed, a new tragic perception arose which demanded the consolation that only art could confer. The nineteenth century solaced its affliction with the symphony and the novel, and it consecrated art by making it the essential ingredient in a religion called 'culture'. The symphony, the novel, and the idea of culture—these were the nineteenth-century consolations. The unprecedented importance that they acquired is explained by a yearning for ecumenical vision; and the unprecedented importance that Venice acquired is explained by its being the most symphonic, novelistic, and cultural of cities.

Schopenhauer anticipated a new way of thinking when he attributed to instrumental music supreme transcendental significance. Such music, he maintained, represented the metaphysics of all that was physical in the world. It was not, like the other arts, an image of phenomena, a copy of the Platonic Idea; but a direct expression of what he called the Will—that is, the noumenal reality that lies behind all appearance. And it was in the symphony that he found the highest form of instrumental music, because

* It hasn't invented very much, this miserable century. It has knocked down everything and raised up nothing.

the symphonic structure resolved chaos and conflict into order and harmony. A symphony of Beethoven revealed the *rerum concordia discors* (the dissonant concord of things), and transmuted human passion into pure abstraction.[11] During the eighteenth century the view had prevailed that instrumental music was inferior. During the nineteenth century Schopenhauer's elevated conception of the symphony was widely proclaimed, and Beethoven, master of the symphonic idiom, was revered. To Berlioz, Beethoven was comparable to Shakespeare. To Wagner, he was a new Luther, the reformer and redeemer of a sacred inheritance.[12] The symphony became a ruling influence in the world of music. A great many symphonies were written, and symphonic thinking overflowed into adjacent areas of composition—into the overture, the tone-poem, and the opera. Especially the opera. Traditional *Singspiel*, in which isolated arias, *arioso*, and ensembles were linked by recitative or dialogue, gave way to the 'through composed' work: a continuous musical sequence in which the orchestra was dominant and in which the *leitmotiv* functioned as an agent of thematic unity. Nineteenth-century opera culminated in the Wagnerian music-drama, which used the voice as a component in an orchestral texture and which Wagner conceived as 'symphonic' in the widest sense. The music-drama was *Gesamtkunstwerk*, a total art-form, that signified the rebirth of Greek tragedy. It recompounded the elements into which the original Attic drama had separated—dance, music, poetry, and the plastic arts. It was Wagner's ideas and Wagner's music that inspired the young Nietzsche to develop his own thesis of dissolution and reintegration. In *Die Geburt der Tragödie* ('The Birth of Tragedy', 1871) music features as the Dionysian art, the art of frenzy and intoxication that is the means of apprehending the original oneness, the ground of being behind phenomena. Such knowledge brings suffering as well as joy, and the Greeks had made it bearable by invoking the arts of dream and illusion—illusion, above all, of individuation. These were the plastic arts, the arts of Apollo. Synthesis had been achieved in Attic tragedy, whose parents were Dionysus and Apollo, the contrasting but conjugal gods. The Greek harmony had been transient; however, Nietzsche's message was that it could be achieved again, now that Dionysus, whose rebirth was signified by modern German music, was overcoming his arch-enemy Socrates.

Thomas Mann linked Venice directly with the Nietzschean idea of redemption through a new cosmic symphony. In his novella *Death in Venice* (1912), the city is stricken with Asiatic cholera. The arrival of the disease, and the grotesque, erotic re-enactment there of the death of Socrates, signal the triumphant return of Dionysus from his Indian exile and his reunion with Apollo in an orgy of intoxication and desire. And other writers responded to its mysterious harmonies, to what Walter Pater would have defined as its aspiration to music. Awareness of *rerum concordia discors* inspired, for example, Gautier's description of the basilica of San Marco:

chose singulière, qui dérange toute idée de proportion, ce ramas de colonnes, de chapiteaux, de bas reliefs, d'émaux, de mosaïques, ce mélange de styles grec, romain, byzantin, arabe, gothique, produisent l'ensemble le plus harmonieux. . . . Ce temple . . . fait de pièces et de morceaux qui se contrarient, enchante et caresse l'œil mieux que ne saurait le faire l'architecture la plus correcte et la plus symétrique: l'unité résulte de la multiplicité.*[13]

Ruskin wrote in similar terms. He interpreted Venice as 'the field of contest between the three pre-eminent architectures of the world', and the Ducal Palace as the expression of their reconciliation in perfect synthesis: 'The Ducal Palace of Venice contains the three elements in exactly equal proportions—the Roman, the Lombard, the Arab. It is the central building of the world.'[14] Ruskin's treatise on Venetian architecture reflects the symphonic quality of his subject. *The Stones of Venice* is in four sections, and it repeatedly invokes the quantity of three—three maritime empires, three pre-eminent architectures, three periods of Venetian history, three volumes. The Beethovian symphony, it will be recalled, has four movements, of which the first is in tripartite, or sonata, form.

The symphony, then, was transcendental. It revealed eternal meaning beyond flux and disintegration in time. The novel too was defined in new and elevated terms. 'On ne peut', wrote Edmond de Goncourt in 1877, 'à l'heure qu'il est, vraiment plus condamner le genre à être l'amusement des jeunes demoiselles en chemins de fer.'†[15] The novel was not transcendental. It was a transcription not of the noumenal but of the phenomenal; it dealt in the currency of the actual and the contingent; it accepted Locke's psychology of memory and his definition of identity as 'consciousness through duration in time'.[16] Yet the novel too was consolatory, because it restored coherence and pattern to life in its spatial dimension. The modern newspaper, of which Bagehot said 'everything is there and everything is disconnected',[17] had its corrective in the modern novel, which supplied

* the strange thing, which upsets all notions of proportion, is that this jumble of columns, capitals, and bas reliefs, of enamel and mosaic; this mixture of Greek, Roman, Byzantine, Arab, and gothic styles, produces the most harmonious totality. . . . This temple, made up of odds and ends which clash with one another, bewitches and caresses the eye better than the most correct and symmetrical architecture could. Unity issues from multiplicity.

† One really cannot, in this day and age, sentence the genre to further existence as the amusement of young ladies in railway carriages.

connections and relationships and put a frame round the whole of human experience. The novel was catholic.

Its practitioners defined it as history. It held a mirror up to life, not to other literature. But it was history of the life that historians did not encounter. The Goncourts, who wrote history as well as novels, stressed the affinity between the two. History was a true novel—'ce romain vrai'—and the novel was 'cette histoire individuelle qui, dans l'Histoire, n'a pas d'historien.'*[18] The modern novel, they explained, 'n'a plus rien de commun avec ce que nos pères entendaient par roman. Le roman actuel se fait avec des documents, racontés ou relevés d'après nature, comme l'histoire se fait avec des documents écrits. Les historiens sont des raconteurs du passé; les romanciers des raconteurs du présent.'†[19] These ideas were imported into English by Henry James. He stressed again and again that the novel was history and that the novelist was a historian who explored 'museums of character and condition unvisited'. His province was 'all life, all feeling, all observation, all vision . . . all experience.' For James, the modern French novelists were not catholic enough. He reproached them for their 'narrow vision of humanity'. They had lost the amplitude of Balzac, whereas the English and Russians had followed where Balzac had led. Trollope's novels referred to 'the whole area of modern vagrancy'; their tone was the 'tone of allusion to many lands and many things'. Turgenev showed the individual 'in the general flood of life, steeped in its relations and contacts'. Tolstoy was 'a reflector vast as a natural lake'. Yet the novel was more than just the sum of documents and observations. The novelist who, like Zola, became a recording mechanism, 'labour[ing] to the end within sight of his notes and charts', dealt only in 'experience by imitation'. The novel took note of science as it took note of everything else; but 'the game of art' had to be played.[20] This idea of the novel as a portrait that was true to life yet at the same time more than life alone was central to French literary theory. 'Ce que la vie lui offre', wrote Flaubert of the novelist, 'il le donne à l'art.'‡ Maupassant defined the novel as 'le miroir des faits, mais un miroir qui les reproduit en leur donnant ce reflet inexprimable, ce je ne sais quoi de presque divin qui est l'art.'§[21]

* that individual history which, in History, has no historian.

† no longer has anything in common with what our fathers understood by 'novel'. The present-day novel is constructed from documentation, either verbal or taken straight from nature, in the way that history is constructed from documents that are written. Historians are narrators of the past; novelists are narrators of the present.

‡ That which life offers him, he offers to art.

§ the mirror of facts, but a mirror which, in reproducing them, gives them that indefinable reflection, that quasi-divine something or other that is art.

All the great cities of the nineteenth century were, in their plenitude of character and incident, novelistic; but to novelists who reckoned that their task began where the historian's ended, modern Venice was especially and compellingly so. Everything about it seemed contrived, to use another phrase of Henry James, for 'putting one in the mood for a story'.[22] Unlike Paris, and London, and Rome, it had moved beyond the province of the historian and become the refuge of those whom history had either forgotten or ignored. The written history of Venice ended with the collapse of the Venetian Republic in 1797; but thereafter its unwritten history had become uniquely enriched. In the later years of the nineteenth century a procession of foreign visitors and settlers, representing the world of privilege and power on vacation or in exile, mingled with the human and architectural remnants of Venetian prestige and transformed the city into an unrivalled museum of character and condition. Modern Venice was territory in which the historian had no mandate. It was territory of the private life; and when the historian tried to reclaim it, he was baffled by the stratagems of secrecy. Those encircling waters were a barrier beyond which the novelist was king. So the novelist returned again and again, inspired by inexhaustible suggestiveness. 'The painter of life and manners', wrote Henry James,

> as he glanced about, could only sigh—as he so frequently has to—over the vision of so much more truth than he can use. What on earth is the need to 'invent', in the midst of tragedy and comedy that never cease? Why, with the subject itself all round, so inimitable, condemn the picture to the silliness of trying not to be aware of it?[23]

The new conception of the symphony was a German response to the nineteenth-century malaise; the new conception of the novel was French; and the new idea of culture was Anglo-Saxon. In French and German thought 'culture' signified a whole way of life. Its scope was national; and it was inside history and determined by it. It could thus be read as an index to social and economic conditions. This was Marx's view, and Taine's. But there were British and American thinkers who understood the term differently. In their view culture was both wider and narrower than Franco-German usage allowed. It was not national, but global; and it signified not the generality, but the best, of art, thought, and manners. Culture, in Matthew Arnold's famous definition, meant 'getting to know, on all matters which most concern us, the best which has been thought and said in the world'.[24] It therefore meant, in practice, knowledge of great art and literature, international travel, and familiarity with foreign languages. Arnold fixed in the Anglo-Saxon mind the notion that culture is an attribute not of the mass but of a minority ('the friends and lovers of culture', 'the poor disparaged followers of culture'), and that it is the antonym of vulgarity. For 'vulgar' too changed its meaning in this general adjustment of terminology. 'Vulgar', which had once meant no more than 'popular', now came to mean coarse, boorish, ignorant, and provincial as well. John Addington Symonds, the essayist and historian whose style of life and writing made him a type of Victorian culture, tried to disown the preciosity and priggishness that the term had come to imply. He acknowledged the justice of Walt Whitman's strictures against excessive concern for sensibility and learning. Nevertheless, his definition of culture remained incontrovertibly Arnoldian: 'It is the appropriation of the heritage bequeathed from previous generations to the needs and cravings of the individual, in his emancipation from "that which binds us all, the common".'[25] Culture, then, in Anglo-Saxon thinking, was the exception, not the rule; and it differed further from the Continental concept in that it was not determined by history, but existed outside history and corrected it. To say that culture consists in knowing 'the best which has been thought and said' is to say that culture is concerned with something that has survived the test of time; with a residue that remains when history has receded. George Eliot stressed this quality of permanence when she wrote of 'that great treasure of knowledge, science, poetry, refinement of thought, feeling, and manners, great memories and the interpretation of great records which is carried on from the minds of one generation to the minds of another'.[26] The belief that this inheritance was therapeutic was taken from Coleridge, who had advocated 'cultivation' as an antidote to the hectic excesses of 'civilization'; and it received its best-known expression in Matthew Arnold's argument that culture was the antidote to anarchy. As T. S. Eliot said, culture in Victorian England was a substitute for religion.[27]

Before the middle of the nineteenth century Venice would not have qualified for inclusion in Arnold's category of the cultural. The British, by and large, did not share Rousseau's indifference. They admired and described the city's canals, bridges, and palaces, and these were constantly reproduced in paintings, engravings, and stage-scenery. Visitors often remarked, like Byron, that they knew the city before they saw it. Yet it remains true that in these earlier years the British did not regard its art and architecture as the best. The gaudy splendour of Venice suggested 'civilization' rather than 'cultivation', and Coleridge might well have had the Republic in mind when he wrote that 'a nation can never be a too cultivated, but may easily be an overcivilised race'. Furthermore Venice was at this time a frontier city. It bordered the Orient, and its civilization carried the stigma of miscegenation. The work of the most prolific and famous Venetian painters was tainted by sensuality. Much Venetian architecture seemed disconcertingly alien in detail and inspiration. It was not until the second half of the nineteenth century that evangelical prejudice was overcome and the mental map redrawn. Venice was now shifted from the frontier between civilization and barbarism to the eminence where civilization and culture intersected. No longer disqualified by their ostentation and their impurity, its art and architecture were reclassified as superlative, and generations of Britons and Americans hungry for culture made the city their Jerusalem. Venice did not command a monopoly of veneration, even when its rehabilitation was complete. Other Italian towns—Florence and Rome most notably—were holy places equally if not more crowded with cultural pilgrims. Furthermore, Rome and Florence were equally well provided with cosmopolitan communities of experts and connoisseurs, ministering like resident priesthoods to the cult of culture. But by reason of both its geographical and its political situation, Venice was better able than they to match the idea of a precious residue, refined by and set apart from the turbid flow of history.

In an age when educated feeling was dominated by the symphony, the novel, and the idea of culture, Venice could not but qualify as a city of the soul, a repository of consolations, a *patrie idéale*. In the work of George Sand, Maurice Barrès, Frederick Rolfe, and John Cowper Powys, it is associated with the androgyne—a much-favoured symbol of wholeness and virginity.[28] This fetishistic perception of Venice was essentially a foreign one, and although its influence was felt in Italy there has been among Italian intellectuals and public figures a determined effort to resist it. Even before the First World War a challenge was issued against the image of Venice abroad, in the name of a modern industrial city that forswore everything the tourists adored. More recently nineteenth-century Venice has been reclaimed for history by Italian historians, who maintain that this period was one of the most dynamic, enterprising, and innovatory that the city has known.[29] The argument is powerful. However, it may yet prove the case that historians of modern Venice, though they may deplore, cannot afford to ignore the Venice that has been celebrated and even invented by an unending succession of literary and artistic devotees.

Venice acquired its new celebrity at a time when Western society was becoming more mobile. Major changes in the technology of transport were making the city accessible to a great many people. But rediscovery was not a consequence of travel. Travel, rather, was a consequence of rediscovery. Venice was put on the itinerary of the sentimental journey by novelists, historians, and apostles of culture. All found there ingredients from which they could concoct remedies for bad dreams and cosmic disorder. To the novelist it was a quarry of plot and character; to the historian, a mine of information about the European past and a clue to the mystery of the fate of empires; to the apostle of culture, a paradigm by which to measure and correct the perversities of contemporary society. By looking at Venice from their respective viewpoints the reasons for its kudos are clearly seen; and it becomes apparent that the chronicles of modern sensibility would be incomplete without reference to this paragon of cities. Yet it is also true that the chronicles of modern Venice would be incomplete without reference to sensibility. There was hidden in these states of mind and acts of the imagination a power to shape events; so to discover the rediscovery of Venice is to be reminded of Pascal's observation about Cleopatra's nose. If it had seemed less comely, the whole world would have changed.

# · PART ONE ·

# *Unwritten History*

## · I ·

# *A Carnival Resumes*

FOR most of the nineteenth century the Venice of literature was remote from contemporary life. It was a city of poetry and historical romance. Byron, Shelley, Fenimore Cooper, and George Sand had claimed it for solitary sufferers who communed with themselves about the vicissitudes of human destiny, and for protagonists in stage-costume who addressed each other in archaic speech. Then in the 1870s the Venice of literature changed. It became crowded with characters who wore modern dress and used modern language; it nurtured relationships and social encounters. In the fiction of Henry James, Gabriele d'Annunzio, Thomas Mann, Frederick Rolfe, and Marcel Proust, as well as in that of less notable writers like William Dean Howells, Henri de Régnier, Anna de Noailles, Arthur Symons, and L. P. Hartley, it features as an annexe to the great cities of Europe and America—a theatre of modern love and death.

This repopulation of the Venice of the imagination reflected a change in actual circumstances. In the time of Byron, Shelley, Cooper, and Sand, Venice was neglected and physically isolated. It was accessible only by sea; and since its political independence and most of its commerce had been extinguished in 1797, the captains and the kings had departed. But then in January 1846 a viaduct across the lagoon was opened and the city was linked by rail to Vicenza. By 1857 the line was complete to Milan, and Venice was brought within the reach of the railway-travelling public of Britain, France, and Germany. After the opening of the Mont Cenis tunnel in 1871 thousands of visitors arrived every year and the 'army of tourists' in Venice became a cliché of travel journalism as well as a mainstay of the local economy. The first Cook's tour to Venice took place in 1867, and soon the conducted party was a familiar sight in the celebrated precincts. In 1881 Henry James saw tour-guides 'lead[ing] their helpless captives through churches and galleries in dense, irresponsible groups'. The transatlantic connection developed rapidly between 1851, when the United States' consul reported the passage of 200 Americans through the city, and 1894, when the English traveller Daniel Pidgeon watched 'an American party of great apparent indistinction put out, day after day, in a gondola ostentatiously flying the star-spangled banner'. Nevertheless, for many years Venice did not share the sort of popularity that cheap travel and organized tourism brought to Switzerland and Paris. It was not a 'playground', and its appeal to travellers with no claim or aspiration to connoisseurship was limited. 'Not only a special taste and a special aptitude', wrote the journalist George Augustus Sala in 1867, 'but a special education . . . are needed before the beauties of Venice can be properly appreciated or her pictorial and architectural wonders enjoyed.' Consequently, the average tourist was 'apt to find Venice slow, and to long for some city where there [were] carriages and theatres and balls and concerts.'

The voluptuaries were finally provided for when the Lido was turned into a beach resort. Commercial development began in the early 1880s, with the building of a boulevard and tramway across the island from the landing stage at Sant' Elizabetta and the introduction of bathing-huts, shops, and *pensioni*. The 1897 edition of Murray's *Handbook for North Italy* recorded that although the Lido was associated with Byron, as the spot where he used to ride and wished to be buried, 'the weird look and feeling of solitude which formerly haunted the place' had disappeared. 'Large *Restaurants*', Murray went on, 'have now been erected in connection with the *Bathing Establishment*, and the place is thronged on summer evenings when the band plays.' By the eve of the First World War there were two luxury hotels (the des Bains and the Excelsior), a rash of summer villas, and all the services and facilities that a chic international clientele required. August and September, traditionally the dead season in the Venetian touristic calendar, had by the early 1900s become as busy as the autumn and spring. In 1886 a correspondent of *The Times* was recommending August to 'the true traveller', as the time when Venice was 'the Venice of the Venetians'. In 1910 the novelist Frederick Rolfe noted that the period from July to October was now the season of 'the salaried classes who take an annual holiday'. No longer was Venice 'slow'. In fact it was decidedly 'fast'. The Lido beach was described by a correspondent of *The Times* as 'a parade-ground for the most daring bathing costumes which the taste of feminine visitors from Austria and Hungary could devise'; and male exhibitionism was rampant too, judging by Edward Hutton's scandalized comments. 'As for the men', he wrote, 'only less appalling in appearance than the women, their costume consists for the most part of a pair of small drawers which would scarcely pass on the loneliest Cornish beach.'[1]

Its characteristic combination of artistic richness, literary associations, and seaside sophistication made Venice supremely modish. Rank, money, and fame flowed there from London, Paris, Moscow, Berlin, Vienna, and New York. This was where wealthy Americans bought their bric-à-brac and polished their manners, and where the European *ancien régime* had its final fling. The captains and the kings returned. Their yachts moored at the Molo; their patronage enriched the hotels. In the spring of 1909 Venice glittered with celebrities—Lord and Lady Rosebery; Lord and Lady Drogheda; Queen Alexandra; the dowager empress of Russia; Princess Victoria of Prussia. Ezra Pound remarked in May 1913 on the beautiful clothes and the swank of the Piazza. 'Venice', he wrote, 'seemed yesterday like one large Carlton Hotel.'[2]

Macroeconomic change, by taking away from Venice much of its remoteness and provinciality, reinforced the effect of local contingencies. In the later decades of the nineteenth century technology transformed the role of the Mediterranean in global affairs. After the building of Continental railways, and the opening of the Suez Canal in 1869, the inland sea again became a great thoroughfare. It carried the traffic between Britain and its empire in North Africa and the East; and Venice, specially favoured by the opening of the Mont Cenis tunnel, found itself in the mainstream of modern communications. In 1872 it captured the India Mail, a major prize in the transit business. The Peninsular & Oriental Company contracted with the Italian government to provide a regular mail service between Venice and Alexandria, and this put Venice on the map as the main port of embarkation for British personnel travelling to India and Egypt. *The Times* wrote of 'the 150 to 200 first-class passengers who used every fortnight to bring so much grist to the mills of the hotels and the shops'. *La Peninsulare*, as it was known, became a familiar feature of local life, and its Indian, Japanese, and Chinese crewmen brought back to Venice something of its old racial heterogeneity and cosmopolitan bustle. In 1892, when the contract expired and the P. & O. transferred its operations to Brindisi, the damage to the local economy was such that the *sindaco*, the Chamber of Commerce, the Provincial Council, and the press all petitioned for a restoration of the British line's privileges. In 1895 the Italian government and the P. & O. signed a new convention which, although it did not bring back the India Mail, established a regular British cargo and passenger service between Venice and Port Said. Thus Venice was linked directly once more with Egypt, Aden, India, Ceylon, China, Japan, Australia, and New Zealand. A Sailors' Institute, under British management, was opened in the 1880s on the Fondamenta Minotto, at the rear of the church of the Tolentini, and the total of 2,385 visits registered in 1890 suggests a significant naval traffic. The result of all these causes was a place of familiar faces and renewed acquaintanceships. 'Venice is full, the hotels overflow, and I meet every hour (that I am out) somebody I know or who knows me,' wrote Henry James in 1894. 'Venice', wrote Lady Radnor in her memoirs, 'was certainly a good place for meeting old friends. Everybody, more or less, passed through . . . on their way home from the Continent, or from Egypt.' The domestic imagery regularly chosen by Henry James to designate the Piazza San Marco ('an immense open-air drawing room', 'an open-air salon', 'a great drawing room, the drawing room of Europe', 'a smooth-floored, blue-roofed chamber of amenity') identified Venice as a venue of fashion and distinction.[3]

Among all these shifting crowds of strangers there were a few who found in Venice not a port of call or a holiday resort, but a home. They came and stayed, or returned again and again, contributing to what Henry James called its 'unwritten history'. They gave a special quality to its social texture, and maintained the illusion of a past regained. In the years of the exiles the city was a part of modern life; but it remained apart from ordinary experience. They were amateurs and connoisseurs of old Venice, who inhabited a world where the clocks beat more slowly, where reckonings were deferred, and where there was always a private gondola at the door. To the historian they seem oddly amphibious creatures, belonging half to a literature created by Venice, and half to a Venice created by literature.

Edmund Flagg, the American consul, wrote in 1853 that 'to one aweary of the world, disappointed, chagrined, sick at heart and soul, there is not a spot in all Italy—in all the world perhaps—which can present such attractions for a few years' residence as Beautiful Venice.'[4] That echoes the literary convention, stretching from Byron and Chateaubriand to Barrès and Henry James, which links Venice with brooding misfits and social and political casualties. But it was a convention that was constantly validated by experience. At the time when Flagg was writing, deposed royalty had already adopted Venice as an asylum. In the magnificent gothic Palazzo Cavalli, on the Grand Canal in the parish of San Vidal, a little court of exiled legitimists paid homage to the comte de Chambord, grandson of Charles X, as Henri V of France. His widowed mother, the duchesse de Berry, now remarried to a penniless Italian count, bought the Palazzo Vendramin-Calergi with all its pictures and furniture in 1844, and lived there—in a style well beyond her means—until she died in 1870. The comte de Montmoulin, pretender to the Spanish throne, was installed in the Palazzo Loredan at San

Vio. 'One met these fallen princes in the squares and streets,' wrote the American author William Dean Howells, 'bowing with distinct courtesy to any that chose to salute them.' A trio of bachelors represented British eccentricity and delinquency. The most conspicuous was Rawdon Brown, who lived in Venice for fifty years. He was the first English historian to use the Venetian archives, and dozens of Victorians, including Ruskin and Prince Leopold, found their way about the city under his guidance. Charles Eliot Norton heard that he had come to Venice on a romantic mission to find the grave of Shakespeare's 'banish'd Norfolk', while Effie Ruskin had some intimation of a great secret sorrow; but no one ever knew the real man and at his own request his papers were destroyed after his death, in 1883. In legend he became the Englishman who arrived in Venice as a youth, fell in love with it, and could never afterwards tear himself away. Robert Browning wrote a poem about the celebrated occasion when he was supposed to have packed his bags to revisit England before he died, taken a farewell look at his beautiful home, and changed his mind:

> Down
> With carpet-bag, and off with valise-straps!
> *Bella Venezia, non ti lascio più!*

'Nor did Brown ever leave her,' added Browning. He was wrong. Brown made at least two trips to London in the 1850s. Nevertheless, the claim was enshrined as fact in the *Dictionary of National Biography*.[5] Edward Leeves and William Bankes were both homosexuals with a partiality for soldiers. Bankes, a friend of Byron, lived in Venice for fourteen years as a fugitive from British justice. He was a rich man—in 1834 he had inherited the estate of Kingston Lacy in Dorset—whose social and political career had been wrecked by sexual indiscretion. In 1841 he jumped bail after being committed (for the second time) on a charge of indecency, and until his death in 1855 he lived abroad, chiefly in Venice. Here he collected and commissioned paintings, *objets d'art*, and furniture, which he sent to England to embellish the ancestral home that he was never again allowed to inhabit.[6] Leeves, older and more shadowy, was a wealthy bachelor from Torrington, in Sussex. All that is known of him is that he was born about 1790, lived in Venice for a number of years in the 1840s, left after the outbreak of revolution in 1848, and then returned in 1850 to nurse the poignant memory of a brief but tragic affair with a young guardsman in London.[7]

The patchy record of these early settlers is a premonition of what was to come. However, the characteristically Venetian expatriate—gregarious, cosmopolitan, and speaking English or French—was a long time coming. At this stage the only foreign language widely spoken in Venice was German, and Anglo-Saxon residents were few and isolated. Charles James, whose father, the novelist G. P. R. James, was British consul in Venice from 1858 until 1860, wrote: 'Within a very short time we had come to know them all, and also to learn that they did not know each other.' When in 1895 the cemetery of San Michele was reorganized and part of the Protestant section moved, it was discovered that only thirty-three British graves had been dug there during the previous fifty-odd years.[8] The Florentine and Roman cemeteries were much fuller of Henry James's exiled casualties—'the deposed, the defeated, the disenchanted, the wounded, or even only the bored'. The truth is that Venice was not yet popular among Europeans and Americans who were compelled or inclined to live abroad. They were deterred by its medical notoriety, its high cost of living, and its social and cultural torpor.

Venice was mistrusted by early Victorian doctors, and it never acquired the therapeutic reputation that drew foreign residential communities to other southern European cities. Its dustless atmosphere and soothing aquatic locomotion were acknowledged as beneficial in cases of inflammatory and feverish illness; but its climate was reckoned too sedative for diseases of debility. This meant that the majority of Victorian invalids abroad were warned to avoid it. They suffered from pulmonary consumption, and this disease, especially in its later stages, was associated with lowered vitality. 'In consequence of this sedative or lowering effect,' wrote Dr Thomas Burgess in 1852, 'phthisical patients, whose systems have been reduced by protracted disease, or are naturally feeble and easily depressed, should not go to Venice.' Dr Robert Scoresby Jackson, writing ten years later, endorsed this verdict. He conceded that Venice might occasionally be recommended 'in the very earliest stages of consumption, and in some chronic inflammatory bronchial affections, as also in cases of full plethoric habit of body, which require[d] simply their hyper-sanguineous tendency averted'; but he added that Venice was 'not highly esteemed by physicians generally and [was] comparatively rarely resorted to by other than sight-seeing travellers, or by invalids for a short while only'.[9]

And if the climate deterred the sick, the canals deterred the healthy. Because the canals carried all the city's sewage, they occasionally stank and were always smelly. Dickens wrote of 'the prevailing Venetian odour of bilge water and an ebb-tide on a weedy shore'. To a public educated according to the miasmatic theory of disease, bad smells were lethal. The idea therefore took a firm hold that Venice was more than usually prone to epidemic infection. After the mid-1860s the miasmatic theory rapidly lost support in professional circles; but the travelling public were slower to be converted to new ideas, and aversion to the canals in Venice intensified as drains and sewers proliferated elsewhere. Having removed the sight and stench of human garbage from their own cities, the British and French grew more intolerant of the exposed 'nuisances' of Italy. In June 1880 an irate correspondent complained to the *Building News* that Venice was 'a city built in the most absolute defiance of all the maxims and precepts of modern sanitation'. He went on: 'The boasted canals are merely open sewers, as anyone traversing them in a gondola and disturbing their filthy depths by the action of oars speedily finds out.' Guy de Maupassant's impression was 'que les ingénieurs facétieux aient fait sauter la voûte de maçonnerie et de pavés qui recouvre ces courants d'eaux malpropres dans toutes les autres villes du monde, pour forcer les habitants à naviguer sur leurs égouts'.* The novelist and travel writer Anne Buckland advised visitors that 'in warm weather a bottle of *eau-de-Cologne* and a pocket handkerchief [were] indispensable in passing through the smaller canals', since these were 'so many open sewers'. One visitor found things just as offensive on the Grand Canal. He wrote to *The Times* in July 1883: 'I never remember the smells so bad at Venice as this summer, and they are even worse on the Grand Canal than elsewhere. This no doubt arises from the constant stirring-up of the water, and consequently also the solid deposit of sewage matter forming the bottom, by the new steamboats.' Cholera reappeared in Egypt in 1883 and soon spread to the Adriatic ports. In Venice it lingered for three years, and there were reports of forty fatalities a day in the summer of 1886. The fact that the epidemic did not on this occasion spread to northern Europe reinforced popular prejudice against miasma, and in the medical profession the conviction revived that in Venice local conditions favoured infection. The sanitary state of Venice was debated at length in the British press and the city's notoriety reached its peak, with the *Lancet* endorsing the opinion that 'Venice invites epidemic disease'.[10]

In the 1820s and 1830s life in Venice, like that in Milan, Rome, and Florence, was cheap for Northerners. 'A Venise', wrote Balzac in 1837, 'il faut si peu d'argent pour vivre.'† This was because the cost of property and accommodation was so low. George Sand wrote of palaces that were for sale at ten or twelve thousand francs (£400–£480), and her claim was confirmed by Rawdon Brown's experience. In 1838 he bought the exquisite, though small and dilapidated, Palazzo Dario on the Grand Canal for £480.[11] Mary

* that mischievous engineers had blown up the vault of masonry and paving that covers the ditchwater in all other cities of the world in order to force the inhabitants to navigate their sewers.

† In Venice you can live on so little.

Shelley, in 1842, found that in Venice it was possible to rent a sumptuous apartment and fill it with antique furniture at very modest cost. 'No-one can spend much money in Venice.'[12] However, from this time prices were rising, outstripping those in Florence and matching those in Rome, which was becoming a very expensive city. According to the economist Nassau Senior, the value of property doubled in Venice between 1840 and 1847; and an anonymous article in the *Quarterly Review* for 1850 suggests that the increase was even steeper:

In 1817 the Barone Galvagna purchased the Savorgnan Palace [in Cannaregio] with a large garden (a rarity in Venice) and with the adjacent building belonging to it, for the sum of 18,000 Austrian livres [£600]. The house contained three vast apartments, the whole in excellent repair, adorned with marble fixtures, gilt and painted ceilings, and gay panels painted by A. Schiavone. In 1846 the proprietor rejected the offer of 300,000 livres [£10,000] which was repeatedly made by the Duke of Bordeaux [i.e. the comte de Chambord]. The Grassi Palace, a much finer house more favourably situated, on the Grand Canal, was sold for 70,000 livres [£2,300] ten years ago; it has several times since changed hands, always at advanced prices, till at the same period of prosperity the sum of 400,000 livres [£13,500] was unhesitatingly rejected.

The cost of renting property matched this rapid inflation. In 1851 John Ruskin and his wife paid £17 a month for a six-roomed apartment on the Grand Canal, whereas only four years previously the Brownings had taken their spacious suite in Casa Guidi in Florence for £26 a year. 'We looked at many other lodgings,' Ruskin told his father, 'but they were either all too small or on the unhealthy side of the Canal, and yet were priced at ten to fifteen pounds per month.' G. P. R. James, when British consul, paid £160 a year for a floor of the Palazzo Foscolo. Prices remained at these prohibitive levels for twenty years. The American consul, William Howells, noted in the early 1860s that in Venice not even ruinous houses were cheap. 'Here', he wrote, 'discomfort and ruin have their price, and the tumbledown is patched up and sold at rates astounding to innocent strangers who come from countries in good repair, where the tumbledown is worth nothing.'[13]

This inflation of property prices was a consequence of the incorporation of Venice into the Austro-Hungarian empire in 1815. 'The wealthy proprietors of Germany', explained the *Quarterly*, 'seemed to delight in speculations which brought them to the beautiful shores of the Adriatic.' Other, equally unwelcome results of this affiliation were political repression and cultural and social inertia.

In 1797 Venice had ceased to be a sovereign republic. It had become a provincial city and for seventy years, save during the brief and tragic interval of the rebellion of 1848–9, it was never free from foreign rule. The Habsburg occupation, which followed a decade of French domination, was as deeply offensive to Anglo-Saxon visitors as it was to George Sand, who described it as 'odieuse et révoltante'.[14] It represented those aspects of statecraft that they most despised—bureaucracy, militarism, police. Early Victorian travel literature is full of anti-Austrian sentiment, and Ruskin described how English ladies on first arriving in Venice invariably began the conversation with the same remark: 'What a dreadful thing it was to be ground under the heel of despotism.'[15] Among the Venetian population itself hostility to Austria was strongest at the lower and middle levels. The lower classes hated conscription and the tobacco tax. The middle classes resented restrictions on economic and political activity. Both consequently sympathized with the liberal and nationalist intelligentsia, who saw the future of their city in an Italian rather than a German context. This is why Venice associated itself with the Risorgimento, the campaign for the liberation and unification of the various Italian states. The situation in Venice was especially distressing to sensitive and liberal-minded foreigners after the suppression of the rebellion in 1849, when the Austrians garrisoned a military force there. For the next seventeen years Venice swarmed with troops, and artillery was kept permanently trained on the Piazzetta from behind a *cancellata* in the arcade of the Ducal Palace. Military defences were thrown up along the Lido to repel a landing of Garibaldi's freedom fighters, and civilians wandered there at their peril. Emily and Ellen Hall, on holiday in Venice in 1860, went for a stroll on the beach and were ordered away at gunpoint by Austrian sentries. An officer explained that the sands were out of bounds because recently some ladies had helped prisoners to escape by sea.[16]

These features of Austrian rule might not in themselves have deterred foreigners from settling. Byron had deplored the Austrian occupation but had nevertheless lived in Venice for three years. Furthermore, British and American expatriates had not been kept away from Rome and Naples by Papal and Bourbon governments, in comparison with which the Habsburg regime was benign. It was also philanthropic and uncorrupt. Consequently, it had its British supporters. Nassau Senior wrote of 'the wickedness and folly of the Venetians in rebelling against the best government they have ever enjoyed'. Edward Leeves and Rawdon Brown both liked Venice because of, not in spite of, the Austrian connection. They regarded this as a guarantee of stability and a safeguard for the artistic heritage. Ruskin's view, likewise, was that Venice was safer in Austrian than in Italian custody.[17] Even its critics were at a loss to evoke a real sense of atrocity. One catalogue of Habsburg abominations, that supplied by George Augustus Sala for the *Daily Telegraph*, strains for drama and attains burlesque:

> Austrian sentries . . . before the Zecca and the royal palace . . . The detestable patrols, whose bayonets were continually, morally speaking, prying over your shoulders, or poking into your loins . . . Gray-coated, bandy-legged Croats, sulking or grinning behind the hideous bars of the *cancellata*, like hyaenas in their dens . . . That aggressive standard of black and yellow . . . Those two murderous field-pieces point[ing] menacingly across the Piazzetta . . . The two monstrous gilt eagles flap[ping] their domineering wings from their twin pedestals in the palace garden . . . The Austrian military band . . . White-tunicked or sky-blue-coated Tedeschi loll[ing] over the tables at Quadri's . . . Skulking gendarmes, with murderous-looking cutlasses stuck in their nasty belts.[18]

More likely to have acted as a disincentive to settlement were the gloom and apathy that pervaded Venice during the later Austrian years. After the revolution it was a city in mourning. Intellectuals boycotted public events, and the aristocracy moved to the mainland. Consequently, *palazzi* were desolate, churches empty, and theatres abandoned to the petty bourgeoisie and artisans. The absence of *cognoscenti* was probably partly responsible for the notorious failure of the two operas of Verdi that were premièred at the Fenice opera-house in the 1850s—*La Traviata* in 1853 and *Simone Boccanegra* in 1857. Effie Ruskin wrote in January 1850 that everything was 'dull and spiritless'. The wealthy Venetian families were refusing official requests and invitations and staying away—though Effie was convinced that they acted less from hatred of the Austrians than from fear of patriotic reprisals. Marshal Radetzky, governor of Lombardy-Venetia, declared that he would give balls and dinners and that if the ladies would not come his officers would waltz together; but Venice refused to be gay. The carnival lapsed; Austrian officers, in the words of Richard Wagner, 'floated about . . . like oil on water'; and cues for public celebration were sullenly ignored. Mrs Newman Hall, a clergyman's wife, in Venice in the late 1850s, noted that when a procession of Austrian royalty passed down the Grand Canal, 'no light appeared from any window, no faces looked from the fair marble balconies, no voice cried God bless them'. Soon the principle theatres were closed (the Fenice shut its doors in 1858 and remained dark for eight years) and *palazzi* along the Grand Canal were showing signs of dereliction. 'There is an air of neglect and ruin about these fine buildings', wrote Georgina Max Müller, 'which is most melancholy. They are not only shut up, but evidently uncared for, the possessors all living away from political causes.' Hippolyte Taine made all this known to the French public, and William Howells warned his American readers that 'the conventional, masquerading, pleasure-loving Venice [had] become as gross a fiction as if . . . it had never existed'. His verdict was that there was 'no greater social dulness, on land or sea, than in contemporary Venice'. One lived there a very lonely life.[19]

In the last three decades of the nineteenth century all these impediments were removed. Social and cultural activity revived; prices fell; and medical opinion modified.

The Austrians withdrew from Lombardy-Venetia in 1866. The province was joined to the new Kingdom of Italy, and Venice became once more a city of fashion and cultural vitality. Theatres reopened, palaces were unshuttered, and emblazoned gondolas skimmed the canals. The remaining noble families returned every winter to their great houses, and from the 1870s until the Great War they resurrected in a series of seasons ever more brilliant the legendary festivity and hospitality of the Venetian Republic. In the 1880s and 1890s titled hostesses presided over a sequence of balls, operas, receptions, and celebrations. The contessa Andriana Marcello, lady-in-waiting to Queen Margherita, was at home every afternoon at five. The contessa Albrizzi organized private theatricals and introduced Gilbert and Sullivan to Venice—though inexplicably she chose to put on *The Mikado* and not *The Gondoliers*. The two contesse Mocenigo, sisters who owned the double Palazzo Mocenigo on the Grand Canal, received every afternoon from six till seven, and then again after the opera, when they provided champagne suppers for guests invited by messengers sent from box to box during the performance. The contessa Annina Morosini, reputedly the most beautiful woman in Italy, led the world of Venetian fashion—until, in April 1904, she scandalized high society by remaining closeted with the German kaiser for an hour and twenty-three minutes. Later she sought solace for lost hopes and lost reputation in the arms of Gabriele d'Annunzio. The duchessa della Grazia, wife of the Italian half-brother of the comte de Chambord and inheritress of the Palazzo Vendramin, had a suite of studios built for the use of resident and visiting artists; and she introduced a new craze into Venice by holding weekly tennis parties in her garden. The contessa Papadopoli, in the Palazzo Coccina Tiepolo on the Grand Canal, specialized in lavish entertainments to mark state occasions and royal visits.[20] These and other aristocratic women revived a way of life that had been interrupted but not forgotten, and at their receptions it was difficult to remember that Venice had ever passed under foreign domination. They served the coffee, wine, ices, syrups, and pastries described in eighteenth-century memoirs. Their suites of rooms with card tables, gilded chairs, brocade-covered sofas, and painted screens reproduced interiors familiar from the pictures of Guardi and Longhi. Laura Ragg, wife of the English chaplain in Venice, described these soirées as an experience of curious but pleasing unreality:

> Leaving one's gondola in the darkness, traversing the dim stone water-entrance lighted by torches or lanterns held by liveried attendants, ascending the great staircase to the *piano nobile* and entering the *sala* where the hostess received her guests, one felt like an actor in a beautifully staged pageant, or asked oneself if this were not a dream produced by an evening reading of an ancient chronicler.[21]

Only the light, the antique light of wax candles, had changed. In their place, fixed in chandeliers and sconces, were electric bulbs. Their glare was tempered by the milky patina of old mirrors and the sheen of damasked surfaces; but they were an unmistakable reminder that Venice had left the age of Guardi and Longhi, and entered that of Fortuny and Sargent.

To its reputation as a precious monument, Venice now added that of a forum of modern music and modern art. The Fenice broadened its repertoire to include French and German works. Operas by Bizet, Massenet, and Debussy, and all of Wagner's principal works except *The Flying Dutchman* were performed there before the First World War. Venice was not as enterprising as Bologna or Turin in the promotion of Wagner's revolutionary music, but until the end of the nineteenth century it was more enterprising than Milan or even Paris. The first Italian production of *Rienzi* was mounted at the Fenice in 1874, and it was in Venice that the Italian public first heard the complete *Ring* tetralogy, which was given by a German touring company in 1883.[22] In 1887 a few notable foreign artists, including Sargent, were represented at a National Exhibition of modern painting in the Giardini Pubblici. Encouraged by its success, the Consiglio Communale decided to inaugurate a biennial exhibition of international art. In April 1895 the first Venice Biennale opened in a specially constructed gallery in the Giardini. Over 200,000 visitors came and Whistler was one of the prize-winners. Within little more than a decade the exhibition's importance was such that foreign academies acquired permanent premises on the site. A Belgian pavilion opened in 1907; British, German, and Hungarian pavilions in 1909. The selection committees of the Biennale were cautious, so the work exhibited was fashionable rather than *avant garde*. It was 1903 before the French Impressionists were shown in any numbers, and British art before the First World War was represented by conservative painters such as Alma-Tadema, Burne-Jones, Herkomer, Holman Hunt, Leighton, Millais, Watts, Brangwyn, Crane, and Sargent. Nevertheless, by Italian standards Venice was advanced. Rodin was given his first Italian showing at the Biennale of 1901; and the Galleria d'Arte Moderna, established by the municipality in the Ca' Pesaro in 1902, exhibited even the work of the Futurists.[23]

The allure of Venice at this time was increased by its relative cheapness. In Italian Venice you did not need to spend a fortune in order to be comfortable, and with a small fortune you could be regal. When the Austrians left, the city's economy was seriously disrupted and the price of property collapsed. It remained low throughout the 1870s and 1880s, and the economic revival, which began in the late 1880s, never brought it up to the levels current in Florence and Rome. Even in its most prosperous pre-1914 years, Venice remained a Cinderella among the large Italian cities. 'A big house here, and especially in this *quartier perdu*,' says Mrs Prest in Henry James's *The Aspern Papers*, 'proves nothing at all; it's perfectly consonant with a state of penury. Dilapidated old *palazzi*, if you'll go out of the way for them, are to be had for five shillings a year.' The record shows that in fact it was possible to live even in the more desirable parts of Venice without spending much. In 1891 the young Romain Rolland learnt that a German acquaintance was paying 4,000 francs (£160) a year for 'un beau palais gothique, et jardin, sur grand canal'—which seemed incredibly cheap when he remembered what her fifth-floor apartment in rue Michelet in Paris was costing his mother. And anything less than a palace could be had for a pittance. In 1881 the painter Luke Fildes rented a spacious flat of seven or eight rooms, opposite the church of San Sebastiano, for 300 lire (£12) a month, including service. In 1886 the artist William Hulton and his half-Italian wife Zina leased a huge newly restored apartment in the Calle della Testa, opposite SS Giovanni e Paolo, for £10 a month. It consisted of fifteen rooms on two floors, plus attics, and five rooms on the ground floor or sea-storey. In the 1890s smaller apartments of two or three rooms on the Riva degli Schiavoni or the Grand Canal cost as little as £2 a month; and as late as 1907 the artist Henry Woods was able to take a long lease on a spacious apartment on the Zattere for £60 a year. These prices were lower than those being paid in Florence and only a fraction of those current in Rome.[24] Furthermore, in Venice someone of limited fortune could not only rent, but own, a substantial residence. The Hultons were given the option of buying their apartment for £2,000—not a small sum, but modest in comparison with the £5,000 paid by Robert Browning at this time for his house in de Vere Gardens, Kensington.

On the Grand Canal Browning's £5,000 would have bought a Renaissance palace. In fact it almost did. In 1885 the poet signed an agreement for the purchase of the Palazzo Manzoni (now known as the Contarini dal Zaffo), on the Canal just below the Accademia. It was a magnificent building and bore the cachet of Ruskin's approval. The verdict in *The Stones of Venice* ('a perfect and very rich example of Byzantine Renaissance; its warm yellow marbles are magnificent') made Browning determined to proceed, even though warned that the Manzoni was on the sunless side of the Canal, was damaged from having been used as a barrack and warehouse, and was apparently structurally unsound. There is no record of the price; but it is valid to assume that Browning's £5,000 of disposable capital would have covered the cost both of purchase and of essential repairs and decoration. Indeed, he could not have afforded more. By Victorian standards he was never more than moderately well-off, and the total value of his estate when he died was just under £17,000. In the event the sale aborted—not from failure to provide the money or from loss of interest, but because the vendor, the marchese Montecuccoli, withdrew, pleading a legal flaw in his title. It seems in fact that Browning was gazumped. The owner's family privately admitted that they wanted to raise the price, and it is known that the *palazzo* was sold.[25] In Venice the cost of accessories, too, was modest. Private transport was probably cheaper than in any other large European city. The standard tariff for a gondolier and gondola was 5 lire a day, or £5–£6 a month, and when not rowing, the gondolier provided general domestic service. It was even cheaper, though often more troublesome, to hire man and craft separately. A fully equipped gondola cost 2 lire a day and a gondolier's services 2–2½ lire. This meant that private transport and general attendance could be had for less than £60 a year, at a time when in Britain it cost £75 a year to maintain a pony and gig, and £207 to keep a carriage and pair.[26]

With the demolition of the miasmatic theory of disease the rehabilitation of Venice was complete. When Koch identified the cholera vibrio in 1884 he in effect identified Venice as a low-risk city. In Venice, drinking water, the chief medium of the vibrio, was exceptionally pure. It was supplied either from deep freshwater wells, which were completely isolated from the canals, or, after the mid-1880s, from subterranean aqueducts fed by artesian wells on the mainland. In Venice, it was now discovered, cholera was a poor person's disease, linked to low standards of personal health and hygiene and to contaminated seafood. 'The well-to-do are free from it, with rare exceptions,' wrote Dr J. A. Menzies from Venice in 1886; 'I have not heard of any case in a hotel here.' It became clear that there was little danger, provided that elementary precautions were taken. These were avoiding shellfish, crustaceans, and *frittura* of whitebait; cooking all vegetables; and washing soft and unpeeled fruit. This last is the precaution that von Aschenbach, in Thomas Mann's *Death in Venice*, neglects. To slake his thirst he buys strawberries from a little fruit shop, eats them as he walks, and pays a terrible price for his indiscretion. Strawberries were especially perilous from having been, in Laura Ragg's words, 'grown on well manured lands and picked and packed by dirty fingers'.[27] The once dreaded smells of Venice were now proclaimed innocent. 'The many odours', Dr Stuart Tidey informed the travelling public in 1899, 'are, I am assured, harmless, being caused by the decomposition by drainage of the sulphates of the salt water into sulphides, than which there are no worse-smelling gasses.' In fact Venice was more, not less sanitary as a result of the discharging of sewage into the canals. Not only were these regularly flushed by the action of the tides, they also contained salt water, which was now discovered to have powerful disinfectant properties. Consequently, wrote Tidey, there was 'a marked absence of diseases due to bad drainage in Venice, especially among foreigners'.[28]

· II ·

# *The New Patricians*

As it qualified as a desirable residence, so Venice attracted evocative expatriates. From precincts of power and wealth the roads of penury, heartbreak, and guilt now led to the marble city. Here privilege in search of consolation joined privilege in search of culture. During the thirty years before the First World War Venice abounded with European and American cosmopolites who filled their leisure with art and bartered ambition for cheap magnificence. And as well as a haven for fallen rank and diminished fortune, it figured as a beckoning muse. Painters, like novelists, now flocked to Venice as they had once flocked to Rome. They frequented the Calcina, an *osteria* on the Zattere, and the Orientale, a café on the Riva between the old State Prisons and the Hotel Danieli. There were dozens of studios in the area of San Trovaso, in Dorsodoro, all supplying tourists and dealers from Britain, France, Germany, and the United States with views made popular by Canaletto and Bonington, as well as with scenes of popular life of the type recently brought into fashion by the Venetian artist Favretto. Luke Fildes and his brother-in-law Henry Woods were two of the best-known purveyors of these pictures of Venetian genre, executed in brilliant colour.

Royal refugees accumulated. Following the Bourbons into Venetian exile came a new generation of losers from the hectic casino of dynastic Europe. They included Princess Darinka, widow of the assassinated Danielo II of Montenegro; Don Carlos, prince of the Spanish Bourbons, whose bid for the Spanish throne had ended in military failure; and Prince Augustin Iturbide, adopted son of the assassinated Emperor Maximilian of Mexico. So it happened that one evening in 1883 Robert Browning, like Voltaire's Candide, found himself dining with dethroned highnesses who had come to Venice for the carnival. 'All deposed and all as gay as grigs,' he reported.[1] Sometime in the 1880s the beautiful Princess Ekaterina Dolgorukaya arrived. Long the mistress of Tsar Alexander II, and mother of three of his children, she had become his morganatic wife only a year before his assassination in 1881. Not all these princely *déshérités* kept up the style of the duchesse de Berry and her two sons. Princess Dolgorukaya lived in deep retirement in a small house with a little walled garden adjacent to the Ponte degli Incurabili on the Zattere. Princess Darinka lived with her daughter, Princess Olga, in a modest apartment near the Frari. 'She had a great deal to bear, I fancy,' wrote John Addington Symonds, 'in her secluded and impecunious life at Venice.' However, when she died, in 1892, the municipality gave her an elaborate official funeral.[2] In October the same year Queen Victoria's eldest daughter, the Empress Frederick of Germany, returned as a widow to the Palazzo Benzon. For a few weeks her small plump figure, dressed always in black, was a familiar sight in the salons and in the Piazza. Expatriate Venice bowed and curtsied and half expected her to stay; but finally she moved on, taking the tragedy of her bereavement, her loneliness, and her redundancy to her palace at Kronberg, west of Berlin.[3] Her compatriots were relieved to see her go. She was a presence too sombre even for the muted light of a Venetian *piano nobile*, and she must have seemed doubly incongruous in the Benzon, once the home of Byron's most vivacious mistress. Besides, they were too full of what Henry James called 'the egotism of their grievances' to allow room for hers.[4]

After Habsburg rule had ended Venice turned away from Austria. Foreign influences and connections became much more noticeably French and British than German. However, not all Austrians turned away from Venice. Some were so fond of it that they made it their principal home. Prominent among them was Prince Friedrich ('Fritz') von Hohenlohe-Waldenburg, who mastered the Venetian dialect and settled happily with a Milanese actress in the tiny Casetta Rossa, built specially for him by the architect Domenico Rupolo on a patch of ground beside the Palazzo Corner. Crammed like a jewel-box with the rarest samples of French eighteenth-century painting, ornaments, and furniture, the Casetta Rossa became a rendezvous for the élite of European art, letters, and politics. Lord Kitchener was a frequent visitor, and the house acquired a special connection with the poet Gabriele d'Annunzio. His legendary love affair with the Italian actress Eleanora Duse was nourished by its atmosphere of intimacy and refinement, and during the Great War it served him as a residence while he played his new role of aviator and soldier.[5] Hohenlohe's sister, Princess Marie von Thurn und Taxis, set up a Venetian *pied-à-terre* in the mezzanine of Palazzo Valmarana, in San Vio, where her guests frequently included Eleanora Duse and the Bohemian poet Rilke. Both these Austrian aristocrats were attuned to the new cultural orientation of the city. He recorded his impressions of it in French; she wrote novels in French as well as poetry in Italian.

The centre of French society was the Palazzo Dario, which was the Venetian venue of Parisian connoisseurs and littérateurs. Here resided and presided two eminent *bas bleus*, divorcees who had met on a cultural pilgrimage to Bayreuth and discovered in each other a new reason for living. One was the tall and languid comtesse Isabelle de la Baume-Pluvinel, who wrote under the name 'Laurent Evrard'. The other was Augustine Bulteau, described by Robert de Montesquiou as 'la Shéhérazade de l'encre bleue' and by Louis de la Salle as 'une sorte de Don Juan cérébrale'. She alternated, according to mood, between two pseudonyms—'Fœmina' and 'Jacques Vontade'; but her closest friends knew her as 'Toche'. Henri de Régnier wrote that she reminded him of the sturdy craft that went up the Grand Canal loaded with beautiful fruit ('ces fortes barques robustes qui remontent le Grand Canal chargées de beaux fruits'), whereas her friend was like the gondolas that glided through the narrow waterways—mysterious and seemingly lost ('ces gondoles qui glissent dans les étroits rii, mystérieuses et comme perdues'). In 1898 Mme de la Baume bought the Dario for herself and Augustine, restored it, and filled it with pictures, books, and antique musical instruments. Guests were put up in an adjacent *pensione*, the single storey of the unfinished Palazzo Venier, which Mme de la Baume booked entire whenever she was in Venice. Day after day, for a few months every year before the death of Mme de la Baume in 1912, Henri de Régnier, Léon Daudet, and the sisters Anna de Noailles and princesse Hélène de Caraman-Chimay, together with other habitués of Mme Bulteau's famous salon in the avenue de Wagram, would dine and talk with painters and collectors in the Dario, and then retire to rheumatismal chambers in the Venier.[6] The other great French name among the Venetian *salonières* was that of the princesse Edmond de Polignac, whose aristocratic title and artistic instincts were supported by commercial wealth. By the last quarter of the nineteenth century advertising had made her maiden name familiar throughout the civilized world, and in Venice it was to be seen plastered across billboards on the Ponte di Rialto. She was Winnaretta Singer, a daughter of the founder of the Singer Sewing Machine Company and inheritor of a considerable share of his immense fortune. Although her father was American, her mother was French, and Winnaretta spent most of her life in Europe, as wife and widow of the homosexual prince Edmond de Polignac and as lover of a constant succession of possessive and neurotic lesbians—including Romaine Brooks, Olga de Meyer, Ethel Smyth, and Violet Trefusis. She spent a lot of money on the encouragement of science, painting, and music, and in 1894 bought the *palazzo* that had almost been bought by Robert Browning—the Manzoni—and turned it into a temple of art and love. After the prince died, in 1901, Winnie de Polignac, as she was known to her friends, continued to pass several months here every year surrounded by her paramours and protégés. Like her husband she was passionately fond of music, and the beneficiaries of her patronage included Fauré, Chabrier, Ravel, Stravinsky, Satie, de Falla, Kurt Weill, Milhaud, and Poulenc. Fauré became a close friend. It was while he was staying with her at the Palazzo Barbaro in 1891 that he began composing his exquisite *Cinq mélodies de Venise*, to texts by Paul Verlaine. The first of these, 'Mandoline', was given its première one night on the lagoon, in a large fishing boat. The talented Mme Duez sang it to the accompaniment of a small ensemble, with Fauré himself at a portable yacht piano.[7]

The British and Americans, of whom there were about fifty by the late 1880s,[8] made up the biggest colony of domiciled and semi-domiciled expatriates. In fact they were numerous enough to attract a small contingent of English-speaking doctors and clergymen. Most had an informed interest, and several developed a professional expertise, in Venetian art and history; but their true role in the life of letters was not so much to write as to be written about. For it was one of the peculiarities of this supremely novelistic city that it did not provide an environment in which novelists or playwrights could thrive.

Authors in Venice were like flowers in a vase. Henry James wrote Venetian stories; yet he never acquired a Venetian home, despite talking often about doing so. He seems to have sensed, like the Anglo-Irish dramatist Augusta Gregory, that it was good in small doses. 'I could not bear much of it', Lady Gregory decided in 1909; 'it is too far from my world.'[9] The American Constance Fletcher, whose pen-name was 'George Fleming', discovered that she could function only by spending long periods elsewhere. She had been brought to Venice at the age of 12 in 1870; but after a precocious literary début (with *A Nile Novel*, in 1876) she broke away. In the 1880s, 1890s, and early 1900s she was more often seen in London. Here she enjoyed the friendship of Oscar Wilde and Henry James, became a regular contributor to the *Times Literary Supplement*, and wrote successfully for the West End stage. When, in 1913, she returned for family reasons to live permanently in Venice, her career virtually came to an end.[10] The English Catholic writer Frederick Rolfe, who called himself 'Baron Corvo', lived in Venice for the last five of his fifty-three years, from 1908 until 1913. The city was a powerful stimulus to his imagination; but he contrived to isolate himself, spurning the advances of the Anglo-Saxon colony and killing their sympathy with invective and insult. From the moment he arrived, with his luggage packed in a laundry basket, until the moment he died, unlacing his boots in an attic of the Palazzo Marcello, Rolfe was one of the most conspicuous characters of the city. In his prolonged destitution he was a universal embarrassment; in his brief affluence a public spectacle. John Cowper Powys saw him crossing the lagoon in

> a floating equipage that resembled the barge of Cleopatra, or perhaps . . . that ship, so often delineated in Greek vase paintings, that carried the great god Dionysus on his triumphant voyage. This . . . gondola . . . was actually covered with the most wonderful skins of leopards and lynxes and it was handled by a Being who might very well have passed for the Faun of Praxiteles.

Rolfe was always thus: a weird figure in the distance. He was addicted to the loneliness of the outcaste, to the friction of alienation, to the pain of self-inflicted wounds. Two ideals dominated his writing—priesthood and friendship; but it was a condition of his creativity that they be unattainable ideals. His imagination fed on fantasy, not on possibility. So in ordering his life to serve his art he deliberately put both priesthood and friendship beyond his reach. He invited hostility, adversity, rejection. He bit the hand that fed him; he rejected the sacraments of grace for a baptism of vice. And in place of the life he had compulsively wrecked, he invented a life of rich and chaste fulfilment. In *Hadrian the Seventh* (1904) he became pope. In the novel he wrote in Venice, *The Desire and Pursuit of the Whole*, he found beatitude in friendship of Platonic perfection. Venice enhanced the extremes of Rolfe's divided psyche. His novel is at once caustic and lambent, cruel and tender. It is a lampoon driven by malice into the surreal. When Rolfe wrote of the British residents he filled his enormous Waterman fountain-pen with gall and transmogrified them into gesticulating puppets and gibbering grotesques. But at the same time it is a lyrical celebration of human beauty and love. The book was too libellous to be published until long after Rolfe was dead; but it was read in manuscript by Ivy van Someren, wife of Dr Ernest van Someren, who as an act of charity had allowed him to stay in their apartment in the Palazzo Mocenigo Corner. The result was predictable and subconsciously desired—what Rolfe called 'the pagan ostracism and Fenian boycott which ruined and is killing me'.[11]

The head of the Anglo-Saxon colony, and chief target of Rolfe's lethal satire, was a woman who was a dethroned queen in spirit if not in rank. This was Enid, Lady Layard, the wife and later widow of Sir Henry Austen Layard. Layard was a Chekhovian character who had slipped from eminence and promise to failure and disrepute. His archaeological investigations in Mesopotamia had brought him early fame. Layard the excavator of Nineveh and Nimrud, and discoverer of Assyrian inscriptions that corroborated the Old Testament, had been 'the man who made the Bible true'.

Later his political activities during the critical years of the Crimean War had seemed to prefigure a dazzling future in public life. Layard the scourge of the incompetent aristocracy had been the rising star of the radical Liberals. But then he had been savaged by Sir Wallis Budge of the British Museum as an ignorant and careless excavator; and after four years as ambassador at Constantinople he had made himself mistrusted in ministerial and royal circles as a rash and bigoted diplomat. Consequently all his high hopes—for the foreign secretaryship, for the embassy at Rome, for a peerage—had been dashed; and in 1883 he had retired to Venice, at the age of 66, with his reputation tarnished and his ambitions unachieved. He feigned indifference to his exclusion from national affairs. 'Had I been put in the House of Lords', he told Lady Gregory, 'I should have considered it my duty to reside in England, which certainly would not have contributed to my health or happiness.' He was happy enough in Venice, writing his memoirs, dabbling in local history, and acting as patriarch and benefactor to the Anglo-Saxon community. But his happiness was often curdled by spleen, and everybody knew that he was a disappointed man.

Layard had had business connections with Venice since 1868, when he had founded the Venice and Murano Glass and Mosaic Company, and he had owned a *palazzo* there, the sixteenth-century Ca' Cappello, since 1874. He now became actively involved in the founding of an English church. In 1888 his company lent a room for Anglican services, and to further the project he collaborated with the bishop of Gibraltar and with Alexander Malcolm, a wealthy Scottish merchant who owned the Palazzo Benzon and a timber business at Longarone. They published a letter in *The Times*, appealing for £2,500 to fund a permanent chaplaincy and place of worship. By the middle of 1889 £900 had been raised, and Layard supplemented this by donating to the English community in Venice a warehouse in Campo San Vio that he had purchased in his wife's name from the Venice and Murano Glass Company. He then appealed for a further £600 to furnish the building and improve its 'very barn-like appearance'. It was dedicated to St George and constituted in 1892 as 'the English Church in Venice'; but conversion and decoration, which were carried out by the artist Henry Woods, were not completed until 1897, three years after Layard's death.[12]

Although Ca' Cappello is one of the smaller palaces on the Grand Canal, Layard and his wife never contrived to make it intimate. Their style of life was public rather than private, and the atmosphere of their Venetian home was more institutional than domestic. Ca' Cappello functioned as a private art gallery and an unofficial embassy. Layard was a connoisseur of Italian painting, and his magnificent collection of fifteenth- and sixteenth-century masters was hung in a suite of rooms whose walls had been specially covered in red, yellow, or green damask. One work, the prize of the collection, stood by itself on an easel in the central *sala*. This was Gentile Bellini's portrait of the Sultan Mahomet II. Enid Layard (née Guest), daughter of an ironmaster but granddaughter of an earl, never acknowledged redundancy and still behaved like an ambassadress. Few individuals of rank or celebrity came to Venice without being accorded the honours of her *palazzo*. Queen Margherita of Italy, the Empress Frederick and her son Kaiser Wilhelm, Queen Alexandra, the dowager empress of Russia, the crown princess of Greece, Lord Kitchener, Robert Browning, John Ruskin, Gabriele d'Annunzio—such were the people that Lady Layard entertained. When Marcel Proust was in Venice, in May 1900, she was in London, and there is no mention of him in her journal for October, when he made his second and final trip to the city. Yet it seems that he too was among her many visitors that autumn, because he introduced the Bellini portrait into *À La Recherche du temps perdu*, as one of Charles Swann's favourite pictures. Certainly it is not difficult to imagine Swann among the guests on one of those special occasions at Ca' Cappello, when the tall hostess would appear bedecked with pearls or emeralds or her famous Assyrian necklace of cuneiform cylinders. Attended by gondoliers in white uniforms and gold sashes, she moved like a sovereign among courtiers and spread a banquet that was truly regal in its meagreness and indigestibility. Lady Layard's Tuesday-evening receptions and Friday *soirées musicales* were red-letter events in the Venetian social calendar, even though her hospitality, which allowed no smoking and insisted on strict punctuality, was too astringent to be popular. The younger set called Ca' Cappello 'the refrigerator'. Laura Ragg, whose husband Lonsdale Ragg was Anglican chaplain in Venice from 1905 until 1909, remembered her as a grand Edwardian lady who was second to none and familiar with all. 'Weekly at the Seamen's Institute', wrote Mrs Ragg, 'she would listen patiently to innumerable ballads and join in the chorus of rather vulgar comic songs; and she valiantly put up with the odour of coarse tobacco clinging to the rooms.' Sometimes she would take her guitar and sing popular songs herself.

Since she was less than half her husband's age, her widowhood was long, and she beguiled it with the Ospedale Cosmopolitano, an infirmary whose official purpose was to provide foreign sailors with medical care from staff speaking their own language. 'A day of great excitement', she wrote in her journal on 28 September 1903:

H. R. H. Princess Christian [of Schleswig-Holstein] formally opened our new International Hospital. Since the night two years ago when I formed the project I have had many difficulties to overcome. I had been at the Italian hospital one day to see some sick English sailors who were burnt by an explosion on a yacht. I found they had been in bed 3 weeks and never once had been washed. I could not rest that night, so I [lit] the electric lamp, got a pencil, and wrote down a plan for making a foreign hospital, which I propose[d] at a committee meeting the next day . . . in the . . . Sailors' Institute. . . . I had two sales of amateur pictures for it in London, and have collected and worked for it, and £1,000 has already been invested.

The hospital was accommodated in an old summer villa of the Cornaro family on the Giudecca, which was leased for £25 a year. Initially it contained only eight beds, but extra wards were added as further premises became available. In 1905 Enid bought the villa, together with an adjacent house, for 21,000 lire (£850), and donated them to the English community as a memorial to her husband. She also paid the £400 required for conversion and repair. There were temperamental doctors as well as touchy bureaucrats to deal with, and she recorded that when she returned from London in the spring of 1903 she 'found members of the committee despondent'. She therefore 'offered to take the whole thing on [her] own shoulders'—an offer that was, according to her own account, gratefully accepted. In *The Desire and Pursuit of the Whole* Frederick Rolfe gave a less heroic slant to her participation. He wrote that the hospital was compelled to rely on Italian medical skill because the English and German doctors would not tolerate her interference in professional matters. He claimed furthermore that the institution was cheated by its purveyors, and he dismissed it as 'a private charity managed and maintained by an irresponsible old lady masked by a committee of doddering nonentities'. It was 'an amateurish hen-roost'. Rolfe's informant was Ernest van Someren, who with his partner Dr Higgins ran a nursing home (called 'the English Hospital') in Campo San Polo. Enid's journal confirms that she disapproved of the regime to which van Someren and the German Dr Werner subjected their patients. 'I went over to the hospital to see the Matron Chaffey', she noted in September 1909. 'She has not many patients just now. Dr Van Someren and Dr W. no longer bring us their patients since the explanations which took place before I left as to their treatment of those they brought in. Dr van S. nearly starved one of them to death.' But van Someren had personal reasons too for disliking Enid Layard. He was a frustrated saviour of souls as well as an offended physician. 'Dr van Someren telephoned he wanted to see me,' runs Enid's journal for 21 September 1906.

He wanted to talk about his religious attitude towards the patients he wants to bring to our hospital in case he closes his own nursing home. He has now a religious mania and thinks he should try and convert people here to his kind of Christianity. I told him we must come to an understanding on the subject as I had given a promise to people here that no one's religion should be tampered with in our hospital, and therefore if he thought right to do so with his patients, I could not admit them. . . . When he began to argue with me as to 'Works' and 'Faith' I told him I was not competent to go into that—that I had given my word and could not go back on it.

Rolfe thrived on the revelations of this resentful and bigoted man. He nurtured an aversion for Lady Layard, which he transferred to the hero of his novel, Nicholas Crabbe:

The more he heard of the proprietor of the infirmary, her wit, her wisdom, her pathos, her pity, her charity, her umbrella, her manufacture of enamelled trinkets for the embellishment of her toad-eaters, her gargantuan appetite for incense and subservience and adulation, so much the more he prayed and schemed and intrigued that by hook or by crook, he might be spared her.

The prayer was not answered, because when Rolfe was in the extremity of destitution and gravely sick with pneumonia, her compassion saved his life. 'Heard that Mr Rolfe was very ill', she noted on 22 April 1910, 'and made all arrangements by telephone to have him taken to the hospital. What a curious irony of fate it is that this strange man, who has tormented me and the people of the hospital with abusive letters and postcards, should now have come to us as a pauper, and almost dying!' Rolfe, transferred on her orders from the public ward to a private room, was briefly contrite during his convalescence. He told the matron, Edith Chaffey, that he now regretted having libelled the English colony in his novel, and promised to make amends in a sequel. But the sequel was never written, and his final tribute to Enid Layard was one of macabre mockery. On the occasion of her funeral, in December 1912, he dressed up in the crimson robes and tessellated hat of a cardinal, and hurled insults at the cortège as it passed.[13]

Adjacent to the hospital on the Giudecca, behind a high surrounding wall, there was a private park: six acres of lawns, flower beds, gravel walks, stone-bordered ponds, vine trellises, hedges, and rose arbours, with an enclosure for a herd of fifteen cattle. It was a pastoral haven in an environment of water and architecture; a touch of Gainsborough in the city of Canaletto. Henri de Régnier and Anna de Noailles introduced it into their novels as a scene of *délassement*, dalliance, and seduction. It was known and loved by Rilke and Eleanora Duse. In Gabriele d'Annunzio's autobiographical novel *Il Fuoco* ('Fire') it is a Garden of Agony, the retreat where La Foscarina, the actress who is Duse, broods on her impending betrayal by the poet Stelio Effrena. To the Anglo-American colony, however, it was a Garden not of Gethsemane but of Eden—and not merely because of the prelapsarian perfection of early summer afternoons, when tea was served there at four o'clock. The park in fact belonged to a couple whose name was Eden. Frederick Eden was a rare example of invalidism in Venice. As the result of an accident in middle age he could not walk, and he had discovered in Venice the perfect city for a cripple. 'One floated in a gondola, without the pain and stress that were . . . exacerbated by Bath chair or carriage. No noise, no flies, no dust.' Later he supplemented his gondola with the first private steam-launch to be seen on the lagoon. In the early 1880s he and his wife Caroline took a long lease on the *entresol* of the Palazzo Barbarigo, whose façade, with its eighteenth-century mosaics, overlooks the lower reaches of the Grand Canal. They also had property on the mainland—a villa and garden in the hills near Belluno, where they passed the hottest months, and a farm and derelict villa on the island of Sant' Erasmo, where fodder was grown for their cattle. Henri de Régnier had this in mind when writing the final scene of his rather insipid melodrama *La Peur de l'amour* (1907). Marcel Renaudier fights his fatal duel with his rival in love, Bernard d'Argimel, in the grounds of Monsieur Ainsworth's deserted villa, which is transferred to Malcontenta on the mainland.[14]

Laura Ragg detected 'some of the selfishness of chronic invalidism' in Frederick Eden. Caroline, however, had 'the nature which delights to spoil and to minister', and the marriage, though childless, was happy. Like the Layards, the Edens found in Venice an antidote for bad luck and matrimonial disparity. But the therapy sometimes failed. Others came and were not comforted or reconciled. Among them was Robert Wiedeman ('Pen') Browning, who brought to Venice one of the most indissoluble of human sadnesses—that of the commonplace child of exceptional parents.

Pen was groomed for greatness and doomed to mediocrity. Only child of the most famous literary couple of the Victorian age, smothered with their affection and loaded with their expectations, he inherited nothing but their name. Nature, as determined to make him ordinary as they were determined to make him outstanding, had denied him their looks, their genius, their charisma; and he suffered cruelly from constant and unkind comparison. He told Mabel Dodge Luhan that he never met people for the first time without seeing them start and hearing them exclaim to themselves: 'What! Is that the son of those two poets?' Such was certainly Lady Gregory's reaction. 'He was pleasant enough, and clever,' she wrote, 'but I was disappointed. He did not look like the child of two poets.' Introduced to the world as a paragon of childish beauty, replete with ringlets, lace petticoats, and cavalier hats, the butterfly had turned inexorably into a grub. The infant sylph had become first a plain youth, and then a plainer man, a target for literary ladies' powers of unflattering description. Edith Cooper ('Michael Field') wrote of his 'beef-red face, baldness, and faunish smirk, combined with the heavy sensitive eyes half caught through their slits'. Mabel Luhan described him as 'a small, red apple of a man with a plump little figure . . . and . . . a round, red, bald head and smooth red cheeks'. She added that 'his eyes were a vivid blue and the eyelids unusually webby and rather mongolian, just like a little snake's'. Freya Stark remembered that he once cut a human shape out of an orange and gave it to her as his portrait. 'I can now hardly distinguish', she wrote, 'in my childish memory, the likeness from the original, except that Pen wore a hard straw boater on top of his general roundness.' At Oxford he had excelled at rowing and billiards, and then left without taking a degree. At the suggestion of Millais he took up art, studied hard in Antwerp and Paris, and developed a line in startling nudes. 'He is clearly an artist born', declared his father; but neither contemporaries nor posterity agreed. His principal sculpture, 'Dryope Fascinated by Apollo in the Form of a Serpent', a six-and-a-half-foot nude contemplating a python, was exhibited at the Grosvenor Gallery in 1884 after being rejected by the Royal Academy. The shocked critic of the *Academy* attacked it as 'a grave error of taste', and it failed to find a purchaser. Only twenty-nine from his output of seventy-two pictures had been sold by the time he died, and most of those had been bought by people anxious to spare his father's feelings. 'One suspects a little practical flattery to the poet in all this', wrote Alfred Domett. Sadly, Pen's was not a case of unrecognized genius. Time, the true judge, upheld the initial verdict. His most ambitious canvas, a life-size historical composition called *The Delivery to the Secular Arm*, was exhibited at the Royal Academy in 1881. It showed an officer of the Inquisition standing over a naked girl chained to a dungeon floor, 'the figure of the latter', according to Domett, 'very powerfully foreshortened'. Bought for £300 by Mrs Clara Bloomfield Moore, a wealthy American friend of Robert Browning, and presented to the Philadelphia Academy of Fine Arts, it was taken down and rolled up in 1944, and has been in storage ever since.

Pen's failure to match his parents' achievement was nowhere more apparent than in his marriage. The marriage of Robert Browning and Elizabeth Barrett, following their legendary courtship and elopement, had been an archetype of chivalry and romance; yet its sequel was a rather sordid match between a man who wanted money and a woman who wanted a name. Pen's bride was Fannie Coddington, a rich heiress from New York who endeared herself to Browning *père* by putting her ambition and her fortune at the disposal of his son. 'Miss C. has spoken to me with the greatest frankness and generosity', Browning told Pen, 'of the means she will have of contributing to your support.' They married in October 1887, when Pen was 38 and she a few years younger. Within three months she had had the first of several miscarriages and begun to cultivate an invalidism in imitation of her famous mother-in-law. Her illness, however, was as much psychological as physiological. She was volatile, quixotic, hysterical; and the sight of Pen's nudes filled her with neurotic jealousy. Her noisy tantrums, with shrieks and threats of suicide, bewildered Pen and embarrassed his guests and friends. His aunt, Sarianna Browning, complained grimly that 'one could hear her down the street'. Enid Layard proclaimed that she was mad.

The reputation of his parents and the recriminations of his wife produced in Pen a sense of inadequacy which he tried to smother with inflated ideas and exaggerated behaviour. He developed a protective reflex of oversize gestures; a compensatory *folie de grandeur*. In him the Browning assertiveness took the form of megalomania tinged with desperation. His pictures became bigger and bigger—six-foot and even ten-foot canvases carrying gallons of paint and a tiny talent. He collected a private menagerie which the American critic William Lyon Phelps described as 'a strange collection of exotic birds, some of which he would hold in his hands while they screamed with ear-splitting screeches and kept their formidable beaks in somewhat dangerous proximity to his eyes'. Voracious for property, he acquired large houses in Asolo and Florence. But it was in Venice that he made the biggest and most hopeless gesture of all. In September 1888, with the help and advice of Alexander Malcolm, he bought the Palazzo Rezzonico—'a proceeding', said Layard, 'which I consider the height of folly'. The Rezzonico, the seventeenth-century masterpiece of Baldassare Longhena on the Grand Canal, was one of the largest baroque palaces not only in Venice but in Italy—and it was in a sorry state. Enid Layard described it as 'a huge waste'. The upper floors had been used as barracks by the Austrians, and evidence of military occupation was still apparent. According to Lady Layard, in one room there was 'a great oven for baking bread for rations, which on account of its great weight Mr Browning is having taken down'.

Pen had been infatuated with Venice ever since his first visit as an adult, in 1885, and Browning's efforts to buy the Manzoni had been made with him in mind. 'I buy it solely for Pen,' he told Dr F. J. Furnivall, 'who is in love with the city beyond anything I could expect, and had set his heart on this particular acquisition.' Pen returned to Venice for three months at the end of 1886, and again a year later on his honeymoon. When he proposed to Fannie he was still grieving over the loss of the Manzoni, and he almost certainly married her in order to satisfy his craving for a home in Venice. The often-repeated assertion that she paid for the Rezzonico does not appear to be true. Pen told the family friend Mrs Bronson that it was bought with money given to him by his father, and there is no reason to dispute this. However, it was Fannie's money that paid for its restoration, and it was Fannie's furniture that filled its vast recesses. After Pen's death she claimed £15,000 from his estate on account of expenditure on the *palazzo*. The claim was disputed by Pen's executors, but there is no doubt that he spent lavishly when refurbishing the dilapidated building. His father reported in December 1888 that he was 'occupied all day long supervising a *posse* of workmen'; and when Browning revisited the palace in the autumn of 1889 he wrote, 'what I left last year as a dingy cavern is now bright and comfortable in all its quantity of rooms'. In 1891 Lord Ronald Gower reported that the Rezzonico was 'the finest private house in the whole of Venice, comfortable, as well as splendid, with a lift and "all modern conveniences"'. Henry James too was impressed. He told his sister: 'What Pen Browning has done here, through his American wife's dollars, with the splendid Rezzonico Palace, transcends description for the beauty and, as Ruskin would say, "wisdom and rightness" of it.' Pen intended the place to be both his home and his palace of art, his Bayreuth. The rooms on the top floor he turned into a gallery for his unsold pictures, and the much maligned 'Dryope' was set up in the central courtyard. He tried to expand his domestic profile to match its immensity. The Browning family pedigree, concocted by his father with more vanity than accuracy, was used to justify a coat of arms and an ostentatious livery in red, gold, and silver for his gondoliers. But the Rezzonico, instead of enhancing Pen, diminished him. In its ornate halls his talent seemed even smaller; his looks commoner; his significance less. 'Pen isn't kingly', wrote James, 'and the *train de vie* remains to be seen. Gondoliers ushering in friends from pensions won't fill it out.' Even people familiar with Venice—Henri de Régnier for example—seemed hardly aware of Pen's existence and believed that the Rezzonico had been bought by Robert Browning; and later the guidebooks registered the Browning connection by describing the *palazzo* as the place not where the son had lived, but where the father had died. Robert Browning's death occurred there in December 1889, and some thought the Rezzonico too big even for that. 'Here was no place where a poet might wish either to live or die,' wrote the English traveller Daniel Pidgeon in 1895. 'The man who was "ever a fighter" might, with more dignity perhaps, have met "the foe" in a more modest retreat, such as Casa Guidi, rather than a corner of the great Rezzonico Palace.' Henry James too, remembering that Browning had been 'the familiar and intimate, almost the confidential poet', regretted that he should have died in this 'beautiful, cold, pompous interior . . . the suggested scene, much rather, of emptier forms and salutations, conventions, and compliments'.

For a few years Pen and Fannie played an active part in the affairs of the Anglo-Saxon colony. Fannie helped to raise money for the English church and to run the Sailors' Institute. However, their marital strife was a common topic of gossip and no one was surprised when they separated. Fannie left Pen in 1893. There was never a formal divorce but, despite Fannie's protestations of love and remorse, and Enid Layard's untiring efforts to bring about a reconciliation, they had little further contact. The Rezzonico was put up for sale and Pen and his aunt Sarianna lived partly in Florence (where she died in 1903) and partly in Asolo (where he died in 1912). In 1895, goaded by her spinster sister Marie, who had always resented the marriage and disliked the Brownings, Fannie demanded the return of her furniture, and the *palazzo* was left empty save for a few memorabilia from Casa Guidi, the famous home of Pen's parents in Florence. Pen lent the Rezzonico to the municipality of Venice in 1896 for a charity ball, which Sarianna described as 'the most magnificent fête given in Venice for more than a century', and Fannie returned to live there for a short while in 1899. But then it was returned to the custody of the dead. Until it was bought by the barone Minerbi in 1906 it was silent, save for the occasional footfall of a visitor come to inspect the Browning relics. In the ownership of Pen it had become what Layard had predicted—'a white elephant'.[15]

Fannie Browning's failure as an American hostess was more than made good by Mrs Bronson and Mrs Curtis, whose attachment to Venice was deep and enduring. Katharine de Kay Bronson ranks among the founders of the Anglo-American colony and she did much to create its characteristic flavour—that distinctive *bouquet* which transatlantic wealth, culture, and sensibility evoke from the faded fabrics of the European past. She more than anyone was responsible for the memory of cosmopolitan company and civilized talk in ancestral salons, of evening light and cigarettes on Venetian balconies, which lies like an incantation behind so much of what Henry James wrote about the Old World. She came to Venice at the age of 42 in 1876, and remained there with her daughter for twenty years, though her husband, who, according to Enid Layard, suffered from 'softening of the brain brought on by dissipation', left for Paris in 1880. Her first receptions were lavish. She rented the sixteenth-century Casa Alvisi, on the Grand Canal opposite the Salute, together with a guest suite in the adjacent Palazzo Giustinian-Recanti, and expatriate society, in the words of Henry James, 'pressed into her rooms'. James tried to resist the vortex of her hospitality. 'The milieu', he wrote, 'was too American.' Nevertheless, he became one of her close friends and recalled the long years of her Venetian residence as 'a sort of legend and boast'. He introduced her into *The Aspern*

*Papers* as Mrs Prest and commemorated her reign at Casa Alvisi in *Italian Hours*. Even closer was her friendship with Robert Browning. This flowered after the death of her husband in 1885, when her own health grew frail and her entertainments became more intimate and discriminating. The celibate life did not suit Browning, even when he was a septuagenarian. He had recently been secretly engaged to the widowed Clara Bloomfield Moore, but she had jilted him after a week. She told Enid Layard that 'she found she could not do it', since she was now madly in love with her protégé, the American inventor John Keely. It was in these circumstances that Browning transferred his affections to Mrs Bronson. A relationship that he began for the sake of Pen, whom he wanted to marry Edith, Mrs Bronson's daughter, he now continued with a view to his own remarriage. This azure-eyed, fatigued, and delicate woman, with her fondness for popular Venetian literature and her philanthropic tenderness, restored to him in old age the desire and sense of potency that another invalid, Elizabeth Barrett, had aroused in him in youth; and in those final autumn holidays in Venice, in that 'last of life for which the first was made', he recapitulated his own legend and boast. He stayed for three months with Mrs Bronson in 1888, and in the autumn of 1889 they were neighbours at Asolo. Here, under her solicitous supervision, he finished his last work, *Asolando*, which he dedicated to her as the tribute of an old man's passion:

> By you stands, and may
> So stand unnoticed till the Judgement Day,
> One who, if once aware that your regard
> Claimed what his heart holds—woke, as from its sward
> The flower, the dormant passion, so to speak—
> Then what a rush of life would startling wreak
> Revenge on your inapprehensive stare . . .

The sense of deep enrichment was mutual; but Katharine Bronson could not respond to Browning's physicality. As he experienced a new rush of life, so she became more ethereal, and she chose to spend her final years in a villa outside Florence, far from the worldly splendours of Venice. Long attracted to the Catholic Church, she became a convert shortly before she died, in 1901.[16]

While the Bronson ménage was too American for Henry James, that of the Curtises was too anti-American. 'They can't keep their hands off their native land, which they loathe,' he wrote to Grace Norton in July 1887, 'and their perpetual digs at it fanned (if a dig can fan) my patriotism to a fever.' Daniel Sargent Curtis enjoyed the reputation of a transatlantic Bayard, banished for his gallantry. According to the British diplomat James Rennell

Rodd, who stayed with him in Venice in 1895, he was 'an old American gentleman who had suffered for his chivalry. He had knocked down a policeman who had been insolent to his wife, and endured the inevitable consequences.' The consequences had been two months in prison, but the causes had been glamourized for European consumption. There had in fact been an argument in a tramcar in Boston and Curtis had been convicted of assault and battery after twisting the nose and breaking the glasses of a judge called Churchill. In 1879, unrepentant and complaining loudly about the uncouthness of America, he brought his British wife Ariana and his eldest son, the artist Ralph Curtis, to Europe. They settled in Venice, in the gothic Palazzo Barbaro on the Grand Canal, and consoled themselves with patrician graces and refinements. 'I felt like my own lady,' wrote Julia Cartwright after visiting them in 1906, 'landing under the pointed arches of Ca' Barbaro and climbing the stairs with flowering oleanders and palms along the steps and bowing gondoliers and Italian servants to welcome me.' They also acquired a garden, the grounds of the Villa Vendramin on the Giudecca, and with these ample social resources pampered passing literary and artistic lions. Robert Browning, Henry James, and John Singer Sargent (who was a relative) became especially well known at the Palazzo Barbaro. It was here that Browning gave readings of his own poetry, charming and surprising his audience with his low-pitched, conversational delivery—'a rebuke', as John Addington Symonds said, 'to the declamatory reading of his poems'. In November 1889 the Curtises and their guests heard the first recitation of poems from the still unpublished *Asolando*—a performance of two hours, with one short interval, which the old man delivered standing. Henry James overcame his initial disapproval and grew quite attached to the Curtises. They were 'a most singular, original, and entertaining couple', 'the soul of benevolence', 'our inimitable Barbarites'; and he stayed with them in 1890 and again in 1907, the year before Daniel Curtis died. He became even fonder of their *palazzo*, with its 'divine old library, where . . . you . . . gazed upward from your couch in the rosy dawn or during the prandial . . . siesta at the medallions and arabesques of the ceiling'; its 'ever adorable . . . marble halls'; and that 'vast, cool, upper floor, all scirocco draughts and easy undressedness' which they had offered to him in 1890 for £40 a year. In *The Wings of the Dove* the Barbaro features as the Palazzo Leporelli, the great house in which the rich Venetian past is revered and served, and which is given up by its owners ('charming people, conscious Venice-lovers') to the dying Milly Theale.[17]

In her autobiography, Zina Hulton remarked on the rarity of Anglo-Italian domesticity in Venice. 'In Florence', she remembered, 'I had been accustomed to so much social life of this kind, resulting from the large number of Anglo-Italian marriages, that I missed it in Venice, where it did not exist.' The foreigners who had come by marriage into Venetian life were almost all Austrian and Greek. She knew of only one other Anglo-Italian couple apart from herself and her English husband. These were the cavaliere and Mme Wiel, who lived in the red gothic Palazzo Pisani at the Ponte delle Erbe in the north-eastern part of the city. Few Anglo-Venetians were as blue-blooded or as blue-stockinged as Althea Wiel. Her father was Lord Wenlock, head of the aristocratic Lawley family of Yorkshire; her mother a daughter of the marquis of Westminster. Taddeo Wiel, her husband, was a native Venetian who worked as a librarian in the Biblioteca Marciana. They had married in 1890, when both were middle-aged, and each had found an intellectual stimulus in the other. It was at her husband's suggestion that Althea wrote her last and most scholarly book, *The Navy of Venice*. But it was not a happy union, and the domestic atmosphere was tense. Taddeo was never fully accepted either by his wife's family or by the Anglo-Saxon colony; and as the result partly of a sense of rejection and partly of an undiagnosed and ultimately fatal cancer, he became neurasthenic and moodily jealous. This in turn increased the intolerance of his wife's friends, who decided that he was a hypochondriac.[18] The only other English wife of an Italian husband was, by the mid-1880s, a widow, and only occasionally seen in Venice. This was the contessa Evelina Pisani, daughter of Dr Julius Millingen, who had been Byron's physician at Missolonghi. Her mother was a French-speaking Levantine who eventually left Millingen, married a Turkish pasha, and published memoirs entitled *Thirty Years in the Harem*. Evelina had married the conte Almoro Pisani, grandson of Doge Alvise Pisani, in 1852, and had dedicated her adult life to rescuing the Pisani estates at Vescovana on the mainland from the general wreck of the family's fortunes. To Henry James, who met her in 1887, she suggested Titian's portrait of Caterina Cornaro, the Venetian queen of Cyprus, or the romantic heroines of Bulwer and Disraeli. 'She is fifty-five and looks forty', he wrote. 'She has spent all her life in Italy; and today, widowed, childless, palaced, villaed, pictured, jewelled, and modified by Venetian society in a kind of mysterious awe, she passes for a great personage and the biggest swell—on the whole—in the place.' Part of the Pisani patrimony was the Palazzo Barbaro, which she first let, then sold, to Daniel and Ariana Curtis.[19]

Anglo-Italian marriages were rarer in Venice than in Florence because the Anglo-Saxon residents were older and fewer. The colony was not quite small enough for everybody to know everybody else, but a diligent hostess would know about half its number. Helen Lady Radnor, who leased an apartment in the Palazzo da Mula for many years, recorded that the twenty-six people she invited to her New Year's Eve party in 1906 were all the English people that she knew in Venice.[20] Althea Lawley's marriage was unpopular not because her husband was Italian but because he was, in the words of Laura Ragg, 'merely a respectable bourgeois'. He had 'neither birth, wealth, nor personal fascination to recommend him'.[21] And if Althea Lawley had married beneath her, Evelina Millingen had done the opposite. She had all the attributes of a low-born schemer and adventuress. The great ladies among the expatriates referred to her scornfully as 'La Turca'. So there is no need to invoke insularity or xenophobia to explain the infrequency of intermarriage or the disapproval aroused by the few cases on record. The Anglo-Saxon expatriates were snobs; but they were neither shy nor contemptuous of things Italian. There was no *ghetto inglese* in Venice as there was in Rome and in Nice, with all the supporting paraphernalia of English shops, English club, and English newspaper. The English newspaper published in Venice, the *Venice Mail*, expired after six issues in 1874; and here even Anglicanism, usually obtrusive and self-conscious when transplanted to the Catholic territory of the Mediterranean, put on a subtle camouflage. The English Church of St George in Campo San Vio, unlike the purpose-built, Gothic-Revival structures in Naples, Rome, Nice, Menton, and elsewhere, harmonized with its surroundings. 'By a happy combination of good fortune and good taste', wrote Lonsdale and Laura Ragg, 'a result has been obtained which suggests at once an English college chapel and the chapel of an old Venetian *scuola* or guild.'[22]

This instance of assimilation typified the Anglo-Venetian experience. The Anglo-Saxons in Venice, unlike their contemporaries in other Italian cities, did not create a world of their own on the margins of local society. They found themselves a role in the world of the Venetians. They became in effect Venetian aristocrats, replenishing a caste that was no longer able to replenish itself, and prolonging beyond political extinction the patrician way of life. They repaired and re-animated decayed *palazzi* and often reclaimed them from commercial or institutional use. They sustained traditions of hospitality, ceremony, and philanthropy; and they patronized the gondolier. Gondoliers had become an endangered species after the building of two iron bridges across the Grand Canal and the introduction of public and private *vaporetti*,[23] and they were probably saved from extinction by foreigners anxious to preserve a traditional and picturesque feature of the Venetian scene. Rolfe accused the English expatriates of exploiting their gondoliers;[24] but there is evidence to suggest that the exploitation was not always on one side. A gondolier could do very well out of service with the

British. Not all were so fortunate as Byron's one-time gondolier, Battista Falcieri, who was brought to England by Disraeli, employed as a valet in the family home, and provided with a government sinecure to ease his old age; but there were plenty of perks and favours to be had if, as was often the case, the employer became emotionally attached or sexually enthralled. For ten years Helen Radnor used every year to bring her gondolier, Giovanni Fasan, to England, where he would row her up and down the Thames in a specially imported gondola. In the summer of 1905 she travelled from Oxford to London in this way.[25] John Addington Symonds, who had a *pied-à-terre* in Venice during the last five years of his life, was inseparable from his gondolier, Angelo Fusato. Ronald Gower, who met them both in Rome in 1891, described Angelo as 'a fine, rough, rather hulky-looking Venetian, who follows him like his shadow'. In 1892 Angelo accompanied Symonds to England, where he went to the music hall, rode on the Brighton omnibus and, dressed in his gondolier costume, was taken on a round of social visits—including one to the Tennysons at Aldworth near Haslemere. This relationship was complex. There was a mixture of obsession and high-mindedness on Symonds's side, and a mixture of naïve devotion and mercenary self-interest on Fusato's. Angelo, like other handsome young Venetians, was making money out of the sexual tastes of a rich foreigner and using this sort of service as a means of escape from domestic drudgery. 'He does enjoy his life on the loose with me', wrote Symonds, 'much more than his life with wife and babies in Venice.' If his attachment to his employer was genuine, the rewards were lasting. After Symonds's death in 1893 the welfare of the increasingly shabby and temperamental gondolier was regarded by Symonds's wife and daughters as a family responsibility, and until he died in 1923 Fusato received a pension of £24 a year, with extra help during the difficult times of the war.[26] Infatuation with gondoliers was a well-recognized characteristic of the Anglo-Venetians, and it earned them more than a few sneers and gibes. Swinburne, in an acerbic essay of 1894, labelled Symonds as 'the Platonic amorist of blue-breeched gondoliers'.[27] Ten years later Arthur Benson, hearing of his brother Fred's social life in Venice with Lady Radnor and the unfrocked priest Charles Williamson, commented in his diary on 'the silliness of it, the idleness, the sentimentality about bronzed gondoliers, etc., with I dare say a nastier background'.[28]

The Venetians were much less disparaging. This is not because they were exceptionally lax or libidinous. There is no evidence to suggest that late in the nineteenth century Venice was still the city of sin that had attracted the debauched and the prurient of all nations in the days of the Republic. In fact it seems to have offered fewer opportunities for illicit sex than Naples,

Rome, or Paris. In Venice homosexuality was especially associated with Germans, and animosity was liable to surface when anti-German feeling was high. This was the case in 1908, when a male brothel on the Fondamenta Nuove was attacked and forced to close. Furthermore, homosexual activity like Rolfe's, which involved youths under the age of consent, attracted verbal if not physical hostility.[29] Nevertheless the demands of Northern European tourists had sustained a limited repertoire of vice and kept clerical opposition at bay. Both heterosexual and homosexual prostitution was tolerated and, provided their liaisons were discreet and legal, British homosexuals were able to live there with little fear of harassment or disapproval. Venice welcomed its Anglo-Saxon residents, whatever their sexual orientation, and the colony became accepted as something authentically Venetian. A writer in the *Gazzetta di Venezia* in 1929 remembered with gratitude how it had enriched local life 'through the creation of individual friendships and intellectual reciprocities'.[30]

# ◦ III ◦

## *Strange Secrets and Broken Fortunes*

ONE of the most famous expatriates of Victorian Venice lived, paradoxically, elsewhere. The ancient American spinster Juliana Bordereau, who features in Henry James's *The Aspern Papers* (1888), hoards in a crumbling Venetian *palazzo* love letters received in her youth from a famous and long-dead poet. However, the original of Miss Bordereau was Claire Clairmont, one of Byron's mistresses, who passed her old age in Florence.[1] James changed her nationality and shifted her from Florence for reasons, he said, of 'delicacy'—to cover his tracks; and he chose Venice as the setting for his story because it was a dramatic requirement that Miss Bordereau's long survival should have gone undetected. It seemed that even a character so remarkable could have passed unnoticed in that city of decayed greatness and mouldy rococo.[2] 'It is a fact', wrote James, 'that almost everyone interesting, appealing, melancholy, memorable, odd, seems at one time or another, after many days and much life, to have gravitated to Venice.' He perceived the city as 'the refuge of endless strange secrets, broken fortunes, and wounded hearts',[3] and in his fiction it became a metaphor for the hidden life.

By now this idea of Venice was familiar. Since Byron had first written of the city in terms of dissolution and oblivion, the theme had been given endless *reprises* by poetasters and travel-writers.

Everybody knew that Venice meant silence and mustiness and extinguished revelry; and Juliana Bordereau, the decrepit Venus who conceals her face behind a mask-like shade, can be read as a personification of what Venice represented to the Romantic imagination. James had pleaded for 'life without rearrangement' in the novel, and he had deplored the 'eternal repetition of a few familiar clichés';[4] yet nothing, at first glance, could seem more contrived and cliché-ridden than *The Aspern Papers*. Here James was apparently transferring characters from the lumber of history to the stage-scenery of romance. In fact, it is in this gothic scenario that his sensitivity to period and place is most acute and his fiction most realistic. The records of the Anglo-Americans who lived in Venice before the First World War show how sound his instincts were when he turned to Venice for 'the illusion of life'. They confirm that the events of his story, while possible elsewhere, in Venice were eminently probable. Indeed, they not only testify to the plausibility of his narrative, they replicate its essential features with uncanny closeness. *The Aspern Papers* mirrored Venice; and Venice mirrored the anxieties of a civilization.

In 1894 James's friend and fellow-novelist, Constance Fenimore Woolson, a lonely and deaf spinster who had spent her life wandering from one European capital to another, committed suicide in Venice by throwing herself from a window; and James preserved her privacy, and probably his own, by taking a bundle of her papers in a gondola across the lagoon and sinking them where the water was deepest.[5] Another of his friends, the novelist Constance Fletcher, who lived with her aged mother in the Palazzo Cappello, the very house that he had in mind when writing his novella, was rumoured to have been the mistress of Lord Lovelace, Byron's grandson, and to have in her possession unpublished letters of the poet.[6] And then there were the circumstances surrounding the papers of John Addington Symonds, who died in 1893. Here we have all the ingredients of a Jamesian *drame intime*: documents detailing a scandalous life locked away for many years in a Venetian *palazzo*; a family's efforts to suppress the truth for the sake of moral reputation and convention; and the inner conflicts of a biographer with divided loyalties and ambivalent motives. James had drawn inspiration from this source in the early 1880s, when he heard from the critic Edmund Gosse about the difficulties of Symonds's marriage. He wrote a rather gruesome short story called *The Author of Beltraffio*, which tells of the tragic conflict between a writer who has a profane view of life, and is governed by the pagan influence of Italy, and his thin-lipped, narrow-minded wife. A short story of 1892, *Sir Dominick Ferrand*, anticipates developments in the Symonds affair by exploring the moral and social consequences which follow when an impoverished writer acquires the compromising papers of a famous dead man. However, James did not comply when it was suggested, after Symonds's death, that he write an appreciation.[7] By this time he knew too much about Symonds's homosexual interests and activities. He knew why 'some of his friends and relations are haunted by a vague malaise';[8] and he had no wish to confront in life the dilemma he had depicted in fiction. There was, he said, a 'strangely morbid and hysterical' side to Symonds's life, and 'to write of him without dealing with it, or at least looking at it, would be an affectation; and yet to deal with it either ironically or explicitly would be a Problem—a problem beyond me'.[9]

So the central figure in this Jamesian scenario was not James himself. It was Symonds's friend and literary executor Horatio Forbes Brown. Brown inherited Symonds's papers, and undertook to write his biography.

Symonds's reputation, which had been considerable, did not long survive his death. He was set aside by the intellectual public as a practitioner of *belles lettres*, whose writing on history and art seemed amateur and out of date in the age of scientific scholarship and formalist aesthetics. But he was not forgotten. After the First World War rumours about his private life, and the known existence in Venice of unpublished papers, tantalized a literary world that had discovered a new genre in frank biography and a special satisfaction in cutting the Victorians down to size. Horatio Brown consequently found his loyalties divided between those who demanded truth and those who demanded reticence; and his uncertainty was increased by the fact that the cause of truth was itself morally ambivalent. Brown, like the writer in James's *Sir Dominick Ferrand*, needed money; and like James's character he knew that where scandal is concerned 'the rectification of history' has the advantage of being lucrative.

Brown was one of those casualties of modern life who were attracted to Venice by its anaesthetizing calm and low cost of living. He was a Scot, proprietor of the small estate of Newhall in Midlothian, and his ambition had always been to combine landowning with literature. But he had discovered that his income was too small for the role of laird, and his talent too slight for that of poet; so he had compromised by settling in Italy and becoming a historian. In 1877 Newhall House had been let; his younger brother had gone to Australia to make his fortune; and Horatio had brought his widowed mother to Venice, James's 'repository of consolations'. In 1881 they took a five-year lease on a large apartment in the Palazzo Balbi-Valier, on the Grand Canal, and in 1885 moved into Ca' Torresella on the Zattere. This was a tenement block of five storeys, with pink stucco façade and white marble balconies, which Brown bought and converted into a handsome *palazzino*, complete with terrace garden. Here he could live in style on modest means, and soon he had become a big fish in the little pool of the Anglo-Saxon colony. He was president of the Cosmopolitan Hospital, treasurer of the Sailors' Institute, and churchwarden of St George's. For twenty years before the First World War Ca' Torresella was a focus for foreign society—especially on Monday evenings, when Horatio held his famous soirées for men. Every week everyone who was anyone among male expatriates and visitors congregated at Ca' Torresella to talk art and literature and pay their respects to old Mrs Brown, daughter of the last of the Macdonells of Glengarry. Even the intrepid Frederick Rolfe cowered before this 'wonderful tottering dame of ninety, in black satin and diamonds', who seemed ready to punish insolence by lashing out with her ebony crutch.[10] And circulating gracefully among the company, a saki bearing wine and whisky, was Antonio Salin, Horatio's divinely handsome gondolier.

In middle age Brown became fussy and donnish, and younger writers poked fun at him. To Ezra Pound he was 'Mrs Horatio'. Leonard Woolf called him 'a caricature pussy-cat littérateur'.[11] But he was no lotus-eater, no *fin de siècle* dilettante. He worked hard and made himself a widely respected authority on Venetian history and Venetian affairs. He knew the archives and he knew the lagoons, and he found a vocation in making the Anglo-American reading public interested in both. His learned works earned him distinguished recognition, including the rank of *cavaliere* from the Italian Crown. But scholarship claimed only half his attention. The other half was devoted to the life and work of Venetian gondoliers, fishermen, and sailors, whose company he enjoyed and whose battles he helped to fight. To Brown, Venice was a social organism—not a museum or a heap of ruins; and he celebrated its human vitality in *Life on the Lagoons*, a little book of enduring interest and fascination. Nevertheless he was a reluctant exile. He fretted about losing touch with friends and literature, and he often ached with longing for home. In 1882 he wrote to his close friend John St Loe Strachey: 'Here I seem buried beyond the reach of all things English and, worse, of all things literary. I sometimes positively hunger for a good bout of some friend—won't you, can't you, come and see us?'[12] He missed things Scottish even more than things English and things literary. He never gave up his membership of the New Club in Edinburgh, and after the death of his mother, in 1909, he often returned to spend his summer in his native territory, living for months in the village inn at Carlops in the Pentland Hills. From there he wrote to his friend and neighbour Lord Rosebery in May 1910: 'I write through the morning and take long walks among these splendidly wild hills in the afternoons, and, not very successfully, try to grow a country air with the help of a Norfolk jacket and a pair of gaiters.'[13] He strove to attract his friends to Venice and to make himself known and remembered in intellectual circles. He stayed with Leslie Stephen at St Ives and with Henry James at Rye. He dined with Edmund Gosse at the House of Lords, with Herbert Fisher at Oxford, and with A. E. Houseman at Cambridge. His correspondence was voluminous. He knew everyone and he went everywhere. But he never knew anyone so well as John Addington Symonds, and he never went anywhere so often as to the Symonds homes in Bristol and Davos.

They had met at Clifton College in 1868, when Brown was a fair-haired 15-year-old schoolboy and Symonds was a visiting lecturer. Although they were divided by only fourteen years, Symonds filled the place of the father that Brown had lost, and Brown filled the place of the son that Symonds never had. Brown chose Venice as his foreign residence partly in order to be nearer Symonds, who was compelled by tuberculosis to move to Switzerland in 1880, and Symonds was for several years tenant of the mezzanine flat in Ca' Torresella. They had few secrets from each other. Symonds advised, admonished, and consoled Horatio in his literary efforts and sentimental infatuations. Horatio knew of Symonds's periodic descents into the sexual underworld of London, Venice, and the big Italian cities. But they did not share the same vices. Brown was no Dorian Gray, no pupil in corruption. Ancestral puritanism, a dominating mother, fear of ostracism and disgrace—all held him back. He stood on the edge—half attracted and half repelled by what he saw; half admiring his friend's excursions into the forbidden reaches of human experience, half dreading their social and psychological consequences. He looked from the world of Henry James into that of André Gide, and he never crossed the frontier. Edmund Gosse was the first English critic to discover Gide, so it is quite likely that Horatio was referring to an early work of Gide (*Les Nourritures terrestres*, perhaps, or *L'Immoraliste*) when he wrote to Gosse in January 1903:

> I've read through this terrible book. The worst of it all is that the analysis is absolutely correct. I've watched it all, and felt some. I was reminded again and again of Johnnie; indeed in some ways the book might stand for a study on him. What a curious spiritual anarchy is produced by this mingling of the intellect with lust. . . . There is a joy of life in the bold discovery that makes the normal and the decent and the measured seem stale and unprofitable. . . . But the end is inevitable—an awful void and a broken will. I am inclined to think the matter may be summed up in one very common word—'health'; though undoubtedly for bold spirits there will always be the query: has man done all there is to be done? Is there not lying just beyond the veil of convention and normality a whole world of knowledge and experience never before tapped?[14]

Symonds died in Rome in April 1893. That summer Horatio went to Davos to take charge of his papers and transport them to Venice. He later told Rosebery: 'The amount of manuscript which came into my possession under his will is indeed immense. Diaries, of the fullest and minutest; an autobiography . . . quantities of unpublished verse; and bundles of letters.'[15] At the request of Symonds's widow Horatio now took on a job that required all his talents of literary alchemy and historical manipulation. He had to turn this full record of an unconventional life into two volumes of conventional biography.

The rules of Victorian biography were clearly laid down. The biographer must let his subject speak for himself; and he must censor closely what he had to say. The dead must not be allowed to harm the living; but neither must they be allowed to harm themselves. The biographer was counsel for the defence. He wrote for an audience obsessed with the notion of sin, and his function was to present the strongest possible case for absolution. He had, as Virginia Woolf put it, to make a funeral effigy and deliver a funeral oration.[16] So he dealt with the public and not the private life. The clothes never came off; the blemishes were kept turned from the light; and the bedroom door stayed shut until the apocalyptic moment of death. The exceptions were few and notorious. Froude's biography of Carlyle, published between 1882 and 1884, provoked outrage because it discussed Carlyle's marital problems. Edmund Gosse's *Father and Son*, of 1907, was variously described as shocking, courageous, and daring because it was less than eulogistic about the author's father. The huge public success of Lytton Strachey's *Eminent Victorians* of 1918 and *Queen Victoria* of 1921 was read as evidence that honesty had finally triumphed over decorum. 'Victorianism', declared Harold Nicolson, 'died in 1921.'[17] In fact the beast was still only mortally wounded, and in its final agony it smothered its assailants. In his biography of Keynes, published in 1950, Sir Roy Harrod wrote about Bloomsbury much as Mrs Gaskell had written about Cranford. During this long ascendancy of hagiography, unsavoury or incriminating evidence was either destroyed or, like radioactive waste, sealed up and buried until time should have reduced it to harmless ashes. Henry James, both in fiction and in reality, had papers burnt. Likewise Lady Burton used fire, with its associations of purification and purgation, to protect her husband's reputation. Evidence of the moral turpitude of Algernon Swinburne and Wilfred Blunt was locked away for fifty years.

Symonds had always chafed against such reticence. 'I say either a truthful biography or none' he wrote in 1868, when Mrs Arthur Clough was consulting him about a memoir of her husband.[18] He once asked Edmund Gosse: 'How long are souls to groan beneath the altar, and poets to eviscerate their offspring, for the sake of what? What shall I call it? An unnatural disnaturing respect for middle-class prudery?'[19] He had seriously contemplated publishing his own studies on sexual psychology, and at the time of his death he was collaborating with Havelock Ellis on a book on sexual inversion. This was to include 'A Problem in Modern Ethics', a pamphlet he had written and had privately printed in 1891. Yet, after his death, he was not allowed to say all that he had to say. Indeed he was not even allowed to say all that Brown was inclined to publish. In this case a double censorship was at work. The censor himself was censored. On his deathbed Symonds gave Brown full discretion to use his papers as he thought best; but he also promised his wife that Brown would publish nothing without her consent. 'I have written things which you would not like to read', he told her, 'but which I have always felt justified and useful for society.'[20] Muffled by this double gag, his voice became all but inaudible.

Horatio was responsive to the advice of nervous friends and obedient to the two commandments of Victorian biography. He wrote to Gosse in January 1894:

> I am going to pack up all diaries, papers etc., and take them to some quiet place in Switzerland and stay there till the work is done. . . . I have settled not to publish the autobiography, just now at least; indeed it would be quite impossible to do so, for various reasons.
>
> The composition of the life and letters will not prove an easy task to me, I imagine. I see how important you think it will be as regards the reputation of our friend, and I also gather that you, and I suppose most of his friends, consider that no allusion should be made, at least directly, to that question which occupied so much of his life and thoughts. I think that if he had lived he would certainly, sooner or later, have opened the whole question in public. But ought I? It would be of great help to me if you would be so kind as to tell me what things have filled you with anxiety recently. . . . I suppose people have been talking, but I trust there is no likelihood of anyone publishing indiscreet letters . . .
>
> I think it will be my duty, even at the cost of veracity, to dwell as much as possible upon the student, the philosopher, the brilliant talker, the almost sophistical dialectician, the genial companion, the religious strain in his nature, rather than upon the actualities of his friendships and his strong desire to be an innovator, even a martyr. However, the solitude of the mountains and the counsels of nature will probably be my best guides through the labyrinth.[21]

His most insistent guide was in fact Symonds's widow, Catharine. Although Henry James never met her, he sketched a recognizable portrait in the character of Mark Ambient's wife, in *The Author of Beltraffio*. Despite fundamental antipathies, Catharine had in fact been tolerant of her husband's work and lifestyle. But, after coping with his chronic illness, with the problems of living abroad, with the death of her eldest daughter, and with all the psychological stress of sexual rejection, she had, not surprisingly, become a dour and withdrawn woman. She was pinched and aged by twenty years of sacrifice and suffering. One of her perpetual worries was the effect on her children of unsettled, expatriate existence. In 1883 she wrote to her sister-in-law Charlotte from Venice:

> There is no need for [the children] to come back to Davos at once even when we get there, if they are having a good time in England. I do want [them] to grow fond of the old country and of my old friends, if they will be so kind as to let them. I have

a morbid horror of their all growing up now into one of those families of expatriated English and Americans whom one meets in places like Venice, with no home or country feeling anywhere, no pride or sense of its being an honour to belong to a noble country. What are art and culture and climate in the limp way in which these cosmopolites take them all to be, compared with the honest sense of *noblesse oblige* which home-feeling gives? That sounds rather like *rant*, but I think you understand what is under it. I meet such lots of these families.[22]

Though she had not repudiated her husband when he was alive, she had no choice but to do so when he was dead. The nature of his relationships and interests made it impossible for her to adopt the role of traditional Victorian widow. He had left no work that she could continue, no cause that she could make her own. In fact she could hope for some remnant of liveable life only by breaking with the past. She therefore decided to move to England and delete from the record all that had made her married life a tragedy. She sold her home in Davos, gave up the mezzanine at Ca' Torresella, and forbade any reference to sex in the biography. In November 1893 she wrote to her husband's old friend, Graham Dakyns: 'The Great Question was supreme in his mind to the very last. Are we right in being cowardly and suppressing it?. . . I trust Horatio fully and want to help, but hinder him as you know.'[23] She found that Horatio, with all his omissions and periphrasis, was still too indiscreet. When she saw the proofs of the book she demanded further cuts,[24] and as finally published early in 1895 it deeply disappointed homosexual readers. Horatio defended himself against the charge of disloyalty to the dead by pleading that the truth was there, between the lines. He wrote to Edward Carpenter, the leading exponent of sexual democracy: 'I have by no means omitted the topic altogether. There are passages on the theory of fellow-service, on the theory of class distinctions etc., which contain the most important of Symonds's views on the subject and which will be understood by those who can understand the matter at all.'[25] But such coded treatment could hardly satisfy Carpenter and his 'Uranian' circle, who had wanted the book to open up debate and foster wider public sympathy.

Horatio had already made it clear to Carpenter that he wanted Symonds's work on sexual inversion to survive. He felt that it would 'provide enlightenment on this difficult but very vital problem' and 'do much for a large and suffering class of mankind'.[26] He dropped the subject from the biography partly because he knew that Havelock Ellis's book was well advanced. This, he was convinced, was a much better vehicle for the topic, which would gain from being discussed in a non-literary context, free from all the decadent associations of art and poetry.[27] His conviction was strengthened by the sordid and sensational trials of Oscar Wilde in 1895, and he encouraged Ellis to reclaim the subject for scientific debate by publishing his book.[28] It appeared first in German, under the title *Das Konträre Gesschlechtsgefühl*, in 1896, and an English edition was scheduled for publication the following year. At this point Catharine again intervened. She asked Horatio to reconsider; and Horatio, in London in the summer of 1897, took further advice—among others from Herbert Asquith, recently Home Secretary. He was told that Symonds's contributions to the book were too literary for the good of the cause, which was better left entirely to medical men; and he reacted by buying up and destroying the entire English edition.[29] Furthermore he vetoed the use of Symonds's name in any future edition. So once more, as a result of loyalty to Catharine, he was accused of disloyalty to Symonds; and when he again made his excuses to Carpenter, in November 1897, he revealed how far he had shifted his ground since 1894:

> I should like to say a word on the charge of having acted unfairly to J.A.S. The question was for me one of great difficulty. I should like to point out that as far as J.A.S.'s place in the history of the controversy is concerned, that is secured by the German book, which contains all he had to say, and more than Ellis was prepared to publish in English.
>
> J.A.S. had all this matter by him for years, most of it in print . . . and yet he never published it. . . . This proves to me that he had at least grave doubts about publishing—of course, in view of his wife and family. He never came a quarter as near publication as I did, and I don't feel sure that he would have faced the inevitable anxiety and possible pain to his family.
>
> You probably do not know that the very last words he wrote, when he was past speech and within a few hours of death, were a strong injunction to me to regard his family in all matters of publication. An appeal from one of his family; the strongly expressed opinion of his oldest and most intimate friends when I got to London; the best legal and medical opinion I could obtain—all combined to make me take the step I did. And though I may not have done quite what he would have liked (but did not do), I think I have done what he would have done in the circumstances.[30]

Whatever the merits of his argument, one thing became clear. His prognostications of pain and anxiety were fully justified. By acting as he did he saved Catharine from a fate more than faintly reminiscent of that of Constance Wilde. No sooner had Oscar been released from jail, than Ellis's book provoked another sensational trial in the cause of public morality. A bookseller who retailed the new edition, published in November 1897 without any attribution to Symonds, was prosecuted and convicted at the Old Bailey of purveying a 'lewd, wicked, bawdy, scandalous and obscene libel'.[31]

Suppressing the truth is a sure way to keeping interest alive; and interest in Symonds survived the collapse of his literary prestige because he was known to be a man about whom the truth had not been told. The possibility that it would be told remained real, because Horatio was not at peace with himself. The charge of disloyalty rankled. He often reproached himself for the way in which he had spoken and written of his dead friend, and he hoped for an opportunity to make amends. After the First World War his inclination to tell the truth grew stronger, partly because there was now a new way of telling it, and partly because he needed money. In the twenty years following the publication of the biography, the science of psychoanalysis revolutionized the understanding of human behaviour and discredited the old moral categories of judgement. Sex replaced religion as the key to the interpretation of character, and, as Virginia Woolf put it, what had once been sin was now misfortune.[32] When Catharine Symonds died in 1913 Horatio was made free to use the papers as he wished, and the surge of interest in Freud and his work after the War convinced him that Symonds's story would now be received with fresh interest and surer sympathy. The climate had changed. The old restraints governing biography seemed redundant, so Symonds could no longer harm himself by saying what he had to say. Furthermore, his research in psychology would give him renewed significance as a Victorian pioneer, a modern before his time. In October 1920 Horatio wrote of his hopes and intentions to Symonds's third daughter, Margaret Vaughan:

> My Dear Madge,
> . . . I always thought and said that your father's position and reputation would have a second blossoming about 20 years after his death and that it was better to wait that time. I have had for the last year or six months a strong instinct that the time has arrived. Letters and talk all suggest this . . . I think the vintage has ripened in the barrels, the sediment gone to the bottom, and it is time to draw off . . . Since I came down from Tirol I have been going through your father's papers—for the third time. I always felt that I must go cautiously, and at long intervals, through the enormous mass, so that I might not, in haste, discard what ought to be kept; and I am writing to Murray to suggest a volume, or two, or more, of letters . . . They thrill me so quickly, they are so alive, that I cannot imagine the world not welcoming them, when I see the fuss it makes over Henry James's letters, which to my mind are such empty wind in comparison (though I loved James himself) . . .[33]

His other motive for reopening the Symonds quarry is clearly apparent in his correspondence with his old friend John St Loe Strachey, who was now editor of the *Spectator*. He told Strachey in August 1921 that he was so hard up that he had to try to make some money by his pen.[34]

Horatio had never been well off, but after his mother's death in 1909 his finances had gone from bad to worse. In 1911, as a result of developing cataracts in both eyes, he had to pay for expensive surgery and nursing care in Zurich. Rents from Newhall were falling, while costs and taxes were rising, and in February 1914 he told Rosebery that he did not know how much longer he would be able to hold on to the estate. He wrote of his dread at the prospect of being forced to part with Newhall, or his home in Venice, or both.[35] A company was floated to exploit the shale deposits at Newhall, but the War both put an end to this and brought special problems of its own. Horatio had to spend the last two years of the War, when it seemed likely that the Austrians would occupy Venice, as a refugee in his own country, living at his club in Edinburgh and in the village inn at Penicuik; and he returned to Venice in 1919 knowing that his worst fears were realized. Ca' Torresella was sold, and he moved as a tenant into the mezzanine flat which he had once let to Symonds. For a time the devaluation of the lira offset the decline in his sterling income; but in post-war Italy prices rose rapidly, as a result of widespread strikes and political instability. Before long he found that he could not afford to leave the country. In August 1920 he wrote to Rosebery from Niederdorf, in the Italian Tyrol:

> I was dreadfully sorry for myself when I found I could not get to England, so sorry that I have not had the heart to write to you . . . but that awful document my half-yearly account settled the [?matter]. I could not afford it, so I came up here to this quiet, cheap little village which I have known for long and always liked, and set myself down to contemplate the inevitable (I fear)—the parting with Newhall. Most of my mother's family, the Glengarrys, died without a roof of their own over their heads, and I doubt whether the blood of 'those beastly Browns' as she called them will save me from a like fate. I am getting over it though, and don't take it too tragically. . . . Doubtless I shall be much better off and freer to come to Scotland when the deed is done.[36]

But Newhall remained without a purchaser for five years; and when it was finally sold, in 1925, Horatio was too ill to travel. He never saw Scotland again.

Old friends from the expatriate colony who came back to Venice after the War found that the hefty, ruddy-faced laird with a common-room flavour of port and ceremony was now infirm, bespectacled, and threadbare—and rather saturnine too. Laura Ragg visited Horatio in May 1920 and saw that he was sadly changed. 'His loosening hold on life was so obvious,' she wrote, 'that I regretted I had not brought our daughter—his godchild—to Venice that she might have said Hail and Farewell to one who had filled so large a place in our Venetian life.'[37] He could no longer offer hospitality or accept invitations. When Asquith's daughter, Princess Bibesco, asked him to a fancy-dress ball, he replied mournfully that he had neither fancy nor dress.

But there was always a dash of self-mockery in his moping, and he still enjoyed sharing a smutty joke with Rosebery. If he had been allowed to disburden himself of the Symonds secret and inherit the profit of his Symonds legacy, he might have died without bitterness.

The text of his new selection of Symonds's letters was sent to John Murray, who had agreed to publish, in the spring of 1922. At the same time, either as a gesture of courtesy or because of some nagging misgiving, Horatio sent a specimen letter to Margaret Vaughan, asking her views about the publication of this and similar material. It was a bad mistake. At once he found himself in conflict with a strong-willed and unhappy woman.

Their feelings for each other had never been warm. Madge had been very close to her father—in Leonard Woolf's view she was a case of the Freudian Electra-complex[38]—and her fixation made her jealous of Horatio, the male cuckoo in the female brood. Horatio, for his part, resented her obsession with her own vulnerability and suffering. Impoverished, homesick, and increasingly lonely, he had little sympathy for a rich married woman who claimed that she faced the world without armour. He once told her: 'Your parents were remarkable persons and had remarkable children who married remarkable men and gave birth to remarkable children and live in palatial drawing rooms and perfect châlets with loads of servants and a French chef. My dear Madge, what thicker armour could you want?'[39] Madge was not only at odds with Horatio; she was also at odds with herself. She wandered between two worlds—the modern world of frank language, sexual liberation, and Post-Impressionism; and the old world of polite conversation, biblical morality, and Academy portraits. Half of her belonged to Bloomsbury. Her closest friends in her early life had been Virginia Woolf and her sister Vanessa Bell—in those days the Misses Stephen. Madge wrote books. She was artistic. She was tempestuous. Men adored her, and she broke their hearts. Her sister Katharine remembered her as 'a meteor passing through life, throwing off sparks and lighting up existence'.[40] The young Virginia Stephen was captivated by her gipsy beauty, her contempt for public opinion, her daring ideas. It seemed her obvious destiny to be a part of that revolution in taste and sensibility that was symbolized by the portentous move of the Stephen daughters from Hyde Park Gate to Bloomsbury. But Madge never arrived where she felt she belonged. She became provincial and respectable. She became a headmaster's wife, and went first to Giggleswick School in Yorkshire, then to Wellington College, and finally to Rugby. Like her father, she touched the imagination of Henry James. Her move to Yorkshire set him musing on the fate of a brilliant young woman banished to a cold and sterile hinterland,[41] and she might easily have become the model of a Jamesian heroine. But in the event it was not Henry James who gave Madge her second existence in fiction. It was Virginia Woolf. Virginia depicted her in *Mrs Dalloway* as Sally Seton: the mercurial, outrageous creature with a vaguely disreputable father and troubled home who ends up in Manchester as the titled wife of a rich industrialist. In the eyes of Mrs Dalloway Sally Seton becomes commonplace, just as Madge had done in the eyes of Virginia. Sally also becomes happy. 'The softness of motherhood, and its egotism too' redeem the platitude of her existence—just as, in Virginia's view, they had redeemed the platitude of Madge's. But to Virginia Woolf no woman could be truly unhappy who had children; and envy made her oblivious of Madge's tribulations.

Madge had married Virginia's cousin, the schoolmaster William Vaughan, in 1898, after agonies of indecision and against the advice of her mother. At one stage she had resolved to become a nun in order to escape from her engagement.[42] Her roles of wife and father's daughter conflicted and made her married life an ordeal. In his public personality Will was 'broad-shouldered, broad-minded, large-hearted'.[43] In private he was egotistical and intolerant, a tyrant with a hearty laugh, who made it his business to tame his wife and train her to serve his own career. He discouraged her literary ambitions, which he considered morbid; and he disapproved of her Bloomsbury friends, whom he considered immoral. In 1920 he forbade Madge and their children to spend the Easter holidays at Charleston, the house in Sussex that Vanessa Bell shared with the painter Duncan Grant. Both Vanessa and Virginia took deep offence, and a lifelong friendship ended.[44] When Madge asked Virginia to review her novel, *A Child of the Alps*, which was published that year, Virginia declined. A few years later her sentiments about Madge found expression in Mrs Ramsay's elegy on dead friendship, in *To the Lighthouse*.[45]

Will's feelings about his father-in-law were close to detestation. He resented the unsavoury rumours and prurient curiosity that surrounded the dead man's memory. These not only offended his deepest prejudice; they also threatened his career. In those days no public-school headmaster could be sure of surviving the revelation of sexual scandal in his family. The strain which all these tensions put on the Vaughans' marriage was obvious to Leonard Woolf, who detected 'the note of matrimonial exasperation' in exchanges between Madge and her husband.[46] It is obvious too in Madge's novel, which is her life rewritten as she wished it to have been. She becomes Linda, half English and half Swiss, half bourgeois and half peasant, a free child of the Alps who is courted by a handsome Swiss *Bauer* called Basil and by a blandly attractive English cousin called Dudley. Uncle Sebastian—a character clearly based on her father—tries to dissuade her from marrying Dudley by warning her of the horrors of middle-class married life 'in some dreary English town or village'. Linda is tormented by conflicting desires; but she finally rejects the simple life of love and nature in Switzerland for the meretricious charms of English respectability and wealth. Thus far the novel is autobiography. It is a reworking of Madge's childhood and adolescent experiences in Davos, Venice, and England, eked out with laboured symbolism. But at the critical point—Linda's marriage with Dudley—it switches from reality to fantasy, and the denouement exposes all Madge's deepest aversions and regrets. Dudley is killed in India; the spirit of the mountains calls; Linda returns to Switzerland; and on the final page she falls into the arms of Basil against a flaming Alpine sunset.

Her book suggests very strongly that Madge felt she had betrayed her father by her marriage; that she needed to atone by allowing him to survive and her husband to die. But if the novel is penitential, it is punitive too. It is the work of a woman who feels not only that she has betrayed her father, but also that her father has betrayed her. It is not clear how far Madge understood Symonds. He had once written to her, 'I want you to know your father'; and he had told her both of his acute sensitivity to the beauty of young men and of his investigations into sexual inversion.[47] In the view of her daughter, Dame Janet Vaughan, Madge was too innocent to put two and two together, too naïve to realize the full implications of her father's addiction to 'Greek love'. Yet there can be no doubt that she came to understand, if only dimly, that her father's sexuality was in some way a threat to her own happiness and a violation of one of the strictest taboos of the world in which she moved. She took her revenge in her novel, where, in effect, she castrates him. Sebastian is an emasculated aesthete, condemned to remain in some profound sense unfulfilled:

> Sebastian became restless; he could not explain what he wanted, he simply felt it; and when . . . he went to stay [in Switzerland] and saw the young men of the village hauling their hay or sitting silent after their day's work—the sunset on their strong limbs, the love-light in their eyes—the craving for his ideal became a literal anguish which stifled him. . . .
>
> He passed through the valleys of mortal anguish. . . . In some ways he was like a woman. . . . He never cared for young ladies. . . .
>
> He was lonely, lonely as all abnormal lovers of the Ideal are lonely. He worked ardently and intensely. He was sure of only one thing in his studies—namely that he was seeking after knowledge. He had only one creed—namely that he must follow good, and thus attain as near as possible to God. He travelled much, for his health was bad. . . . When at home, he read and wrote in a sort of attic-room hung with photographs of Michael Angelo and the young Antinous. . . .

Class was nothing to Sebastian; he loved a peasant just as he loved an aristocrat. Pretensions were what he abhorred, and every form of insincerity. . . . What puzzled Linda had long been outlived by her uncle, in whose frail and battered body little of the mere animal man was left. The spirit was paramount—by that he was guided. He was no saint, but it was impossible for him to be a great sinner.

The letter of her father's which Horatio submitted for her comments prior to publication is lost; but it seems that she found it far too suggestive of both animality and sin. Horatio replied to her protest on 24 April 1922:

I am very sorry that you should feel this anxiety and alarm; and I cannot help feeling that it is excessive. But you are in Rugby, England, and I am in Venice, Italy.

I have given immense pains and attention to the selection, and have gone over and over again the letters I thought might be published. Of course, in such an intimate correspondence as that with Henry Sidgwick and myself it was inevitable that some picture of the man's mind, some anima figura, should be revealed. If that is to be eliminated, there would remain much purely literary matter . . . and a good deal of historical and literary criticism. But whether this would make a book worth publishing I don't know. . . .

If your father is going to have a place in English literature people will always be interested and curious about the man himself. But I understand your anxiety, and I asked your feelings and opinion about a characteristic passage. [I] have got them and will bear them in mind.[48]

This did not satisfy Madge. She turned for help to her brother-in-law, the Classical and Oriental scholar Walter Leaf. Leaf was the husband of Symonds's second daughter, Charlotte. Authoritative, self-assured, and senior in years, he was the family's trouble-shooter. Like Will, he had a professional position to defend, and was terrified of scandal. He was a director of Westminster Bank, a governor of Marlborough College, and a governor of Harrow School. Losing no time, he confronted the publisher, John Murray himself. Murray was conciliatory. The book was not yet in type and there would be no difficulty about removing objectionable material. He had no wish to publish anything that would cause unpleasantness, and would write to Brown. Greatly relieved, Madge jotted a comment in the margin of Leaf's report: 'This letter from Walter Leaf refers to the publication by H. F. Brown of Father's Selected Letters. He had sent me a most disagreeable specimen, which he proposed to publish, and I appealed to the Leafs. The bulky correspondence on this subject I have now destroyed.'[49]

Faced with the combined opposition of his publisher and the Symonds family, Horatio had no choice but to capitulate. So Symonds was muzzled for the third time. The book as it finally appeared in 1923 contained only a passing editorial reference to his interest in psychoanalysis.[50] Madge objected even to that, forcing Horatio on to the defensive yet again:

I myself don't much care for the words you dislike, but I used them on purpose and with a definite object. Your father was much occupied with psychopathology and wrote a lot about it; he often said that he was coming to consider everybody in the light of 'a case', himself included. Had he lived he would certainly have taken an interest in this newfangled, fashionable, Freudian psychology. I meant to indicate that, and it seems I have done so from the letters I get.[51]

The book did better than he expected. A first printing of 1,500 copies sold out within a few weeks and a second impression was brought out in the middle of the year. That too sold well. Almost 1,700 more copies had gone by the end of 1923, and Horatio at last had some spare cash. He told Rosebery that he had been able to buy himself a new suit, a pair of boots, and some underclothes.[52] Nevertheless both he and Madge were dissatisfied. For him the book did not go far enough. It left the central truth untold and its commercial and critical success only served to prove that it could have been even more popular and successful. For Madge it went too far. It set people talking; revived old rumours; raised awkward questions. Arthur Symons, writing in the *Fortnightly Review* in 1924, asserted that John Addington Symonds had been 'physically very sexual . . . [and] to a certain extent abnormal'; and he revealed that Symonds had once told him that the complete autobiography could only be privately printed. He suggested too that Symonds had not been completely sane. Physical sexuality, abnormality, insanity—they all added up to the traditional conception of homosexuality, and even if Madge could not decipher the code she could hardly miss the implication of monstrosity. Her father was undergoing a transformation ever more hideous, and she made a last-ditch attempt to arrest the change and embalm his moral reputation. She wrote her memoirs—an account of her early experiences which proves that Virginia Woolf was right when she accused Madge of never having grown up, of talking a great deal about life but not facing it.[53] Madge wrote in a mood of defiance—defiance of Horatio, of Bloomsbury, of everyone and everything that destroyed innocence and illusion:

I have attempted to write with love, rather than in the spirit of criticism with which of late years it has become fashionable to approach the period which I describe. My parents, who themselves essentially belonged to that period, emerged from it as surely as ever the Phoenix did from his ashes. (And what a gorgeous ash-heap was that of the Victorians!) . . . In writing this 'family record' . . . a strong and reverent affection has been my guide.[54]

She was a sick woman when she wrote her book, and she underwent surgery in the spring of 1925, shortly after its publication. She died in November of that year, aged 56. Her last months were soured by bickering with Horatio, who made carping criticisms of her memoirs and disputed her right to certain of her father's papers. There was a bizarre twist in their relationship in its final stages. The elderly invalid in Venice and the middle-aged invalid in Rugby switched their roles. Horatio now became the neurotic censor, the obsessive suppressor of evidence; Madge became its aggrieved custodian.

Horatio had wanted the truth to be told; but he had wanted it to be told by himself, in his own way and over his own name. He was as jealous, and in his fashion as prudish as Madge in all matters concerning Symonds. He saw himself as uniquely qualified to tell the full story. It was a commission with which no one else could be trusted. In March 1923, shortly after the new volume of letters had appeared, he told Rosebery that this was probably the last the world would hear of Symonds for a long time: 'As I feel at present, I do not think I shall publish any more—or at any rate not much. In case of my death I have left instructions for all my papers to be burned—it is better so. I would do it now but I have a dread of destroying evidence which might be useful in case of attack.' The only exception was to be Symonds's autobiography, which Symonds himself had specially wished to be preserved. Horatio, at the suggestion of his friend Ernest Saltmarshe, intended to bequeath this to the British Museum, with instructions that it was to be sealed for fifty or even a hundred years.[55] However, as a result of inquiries made by Madge during the preparation of her memoirs, further letters had come to light, and it alarmed Horatio to think that these might escape his edict of destruction. From a batch that Madge had sent to him, he knew that they were highly confidential and included letters written by himself and some of his close friends.[56] In December 1924 he asked her to send him the rest of this material and to acknowledge his right, under the terms of her father's will, to deal with it as he thought best.[57] When Madge demurred, he asked her to destroy it:

> I wish you would really think over and answer this question of the property in your father's letters and literary papers. . . . Personally, for myself, I don't mind now what people know about me. I have no preferment, honours, advancement, emoluments in prospect. Every one I care about knows all about me and nothing would make any odds. . . . I . . . don't mind (though I don't desire) publicity, provided it comes through myself. As I say, all this hardly affects me. But I must reflect that there are others who would not feel so free in England, and if anything likely to cause them annoyance should take wing, then if the papers are mine, as I hold them to be . . . I am responsible to them for any divulgence. Think it over . . . but if you come to the

conclusion that I am the owner under your father's will of all his letters, manuscripts, printed books, literary writings, etc., which he did not give to you *himself* in person before his death, then I would ask you to sort out all such papers . . . and burn them and write me a note to say that *at my request* you have burned such papers. That would cover both of us; and I have enough here for any purpose of veracious biography on the right occasion.[58]

Madge's lawyers assured her that Horatio had no legal claim on manuscripts in her possession. It was becoming clear that his intention was an indiscriminate holocaust of Symonds's papers, and they advised her to deposit everything she wanted to preserve in the British Museum. In the event, her papers were donated by her daughter to Bristol University; but the letters that Madge had so unguardedly sent to Horatio were never recovered.

Their relationship ended in mutual mistrust and petty recrimination. Horatio wrote to her for the last time in April 1925, exasperated by her misunderstanding of her father, frustrated in his own efforts to put the record straight, and disgusted by the hypocrisy of people who were accusing him of moral failure:

I am sorry you thought my letter neither kind nor generous. I have now read it through and wonder whether it is really unkinder or less generous than yours in which you told me my friend's sister was one of the most dangerous gossips in Rugby. Anyhow we seem to get wrong when we write to each other on the subject of your father. I am sorry. I think the best plan is to cease writing. I cannot agree with you in your view or picture of him. I never could abide Sebastian; and the two essential psychological passages in your new book don't seem to me to explain anything intelligibly. . . . I wonder if you have any idea of the pressure brought to bear on me to be 'loyal to the dead' by those who knew—that is to say a scattered company of PIGS—what was chiefly occupying your father's mind in the latter years of his life and who looked to him as a pioneer and leader. I shall say no more. It is all very difficult, only pray consider that it may have more respects than one only.[59]

There is something Jamesian about that. It recalls the outburst of Miss Bordereau against 'publishing scoundrels' at the end of *The Aspern Papers*; and Henry James might have written the sequel too. After Horatio Brown died, in August 1926, the Symonds papers, like the Aspern papers after the death of Miss Bordereau, were burnt. A codicil to Horatio's will, dated 6 October 1925, made clear his wish that all literary papers in his possession, save only Symonds's autobiography and a few famous autographs, should be destroyed. The autobiography, initially destined for the British Museum, he had in the event bequeathed to the London Library, with an embargo against publication of fifty years.[60] Following his own recommendation, his trustees and executors, who were his Edinburgh solicitors, consulted Dr

Charles Hagberg Wright, Librarian of the London Library, about the disposal of his papers; and Hagberg Wright in his turn consulted Edmund Gosse.[61] It was Gosse who told Madge's daughter, Janet Vaughan, at tea one afternoon in the late 1920s, what happened next. 'Hagberg Wright and I had a bonfire in the garden and burnt them all, my dear Janet—all except his autobiography, which we have deposited in the London Library not to be available or published for 50 years. I am sure that you will agree that this was the right and proper thing to do.' Janet, a medical student who had inherited her mother's Bloomsbury friendships and sympathies, was furious.

I said very little. It was not safe to let myself speak as I thought of those two old men destroying, one could only guess, all the case histories and basic studies of sexual inversion that J.A.S. is known to have made, together with other letters and papers that would have thrown much light on J.A.S.'s work and friendships. Gosse's smug gloating delight as he told me, the sense that he had enjoyed to the full the honour fate had given him, was nauseating. There was nothing to be said. I walked out and never went back.[62]

With the destruction of Brown's archive were lost not only many details of Symonds's life but a unique record of that network of homosexual relationships and sympathies whose existence, while so difficult to prove, is yet crucially significant in the history of Britain from the mid-Victorian age until the 1930s. We have it on the authority of Havelock Ellis that Symonds 'had come into close contact with many more or less distinguished inverts',[63] and his papers would undoubtedly have made clear how far the male élite that criminalized homosexuality was itself homosexual. They might have said much therefore not only about the psychology of inversion, but also about the psychology of persecution. Symonds's autobiography, which survived, was sufficiently explicit to have aroused repugnance in Hagberg Wright. In 1939 he told Dame Katharine Furse, Symonds's youngest daughter, that he would like to have it destroyed, as 'not conducing to add to the reputation of an author whose works I have read and admired and bought'.[64] However, its historical value is small. It is in essence a retrospective analysis by Symonds of his early psychological and religious development. It has comparatively little to say about his adult life and work. Brown thought that it gave a distorted picture of the man. He told Rosebery that it was 'written too late in life and too much under one impulse to be either true or publishable'.[65] The attitude of Katharine Furse indicates that there was a lot more to Symonds's story than it contains, and that to those with a fuller knowledge it was nothing very shocking.

Katharine was the most notable of Symonds's daughters. Widow of the painter Charles Furse, she had a distinguished record of war service and was for ten years from 1928 director of the World Association of Girl Guides and Girl Scouts. In 1939 she advertised in the press for information about her father. She was writing her memoirs and wanted to find out which of his papers had been destroyed and why. On 31 July she wrote to Virginia Woolf:

> There is indeed 'a mine' in which to work, and I am getting letters from Australia and America, full of interest, following on some publicity about my writing a memoir of Father.
>
> Whatever Macmillan do or do not do, I want to write a book about Father later on and will show it to the Hogarth Press because I feel we might be more in sympathy.[66]

Virginia was encouraging. She urged Katharine to 'let the cat out of the bag', and at first Katharine was inclined to comply. She waved aside counsels of prudence and reticence. On 26 October she told Virginia:

> It is rather pathetic the way some men have been writing to me. . . . If only I were better able to help in the situation. Others have begged me not to write of the subject, implying that I risk my own good name, which does not interest me; but also implying that it may react on the G[irl] G[uide]s. Ridiculous, but such is prudery.[67]

However, Katharine felt much less secure than she would admit. She was herself involved in a lesbian relationship, and as a fuller picture of her father's sexual life emerged, she decided that discretion was, after all, the better part of valour. She wrote to Hagberg Wright on 23 December 1939: 'I am very anxious to complete my records, and not for publication, as I know so much now that I realize there is much to be left unsaid.'[68] And in her memoirs, *Hearts and Pomegranates*, which were published in 1940, she left so much unsaid that she felt obliged to apologize to Virginia:

> What you wrote about urging me to let the cat out of the bag is not in my text. I re-wrote the chapter about J.A.S. from that point of view several times and each time shed some of my inhibitions. You may detect some still. One is haunted by the feeling that whatever one does and writes in this sort of direction may affect one's chances of helping boys and girls. A woman once wrote to Baden Powell that I was so immoral that I ought not to be allowed to do anything [for] the country's children.[69]

What is interesting about this loss of nerve is the fact that it had nothing to do with the autobiography, which she had not yet read. The London Library Committee, at the insistence of Desmond MacCarthy, Harold Nicolson, and E. M. Forster, had finally agreed to her request to read the manuscript in October 1939;[70] but it was another ten years before she did so. The delay is of course partly explained by the outbreak of war; but it seems very odd that she should not have made a special effort to look at such a cardinal document before she published her memoirs—especially as she had been agitating for the right to do so since 1927. We can only surmise that what she had learnt in the mean time had made her very apprehensive as to what she might find. If so, her fears were groundless. When she did finally read her father's autobiography, in 1949, she found very little that she wanted to suppress. She suggested to Lord Ilchester, chairman of the Library Committee, that it might be 'lightly edited for the sake of a few other people';[71] but when Ilchester declined to have it tampered with, she did not insist. In fact she let it be known that she saw no reason to maintain the embargo. 'With some difficulty I got permission to read it,' she told Professor Anthony Blunt in October 1950, 'and much wish that it was not locked up in a brown paper parcel in a cupboard.'[72] She died in 1952, without having published any study of her father; and either before or after her death the spirit of the censor made its final and irreparable descent. None of the evidence she had collected survives among her papers, and the truth is now beyond recovery.

So the episode of the Symonds papers ends in truly Jamesian fashion, with a gesture that completes its symmetry and a denouement that leaves mystery intact. For, like *The Aspern Papers*, this chapter of unwritten history contains not one story but two, with Venice as the element that is common to both. There is the story of the papers; and there is the story in the papers. The story of the papers takes place either in Venice or with Venice very much in mind. It is the story that the historian, in life as in the fiction, is able to tell. The story in the papers has Venice, the city of masks and silence, as its symbol. It is the secret from which the historian is barred. In the novella he is thwarted by the novelist's contrivance; in life, by events which echo and imitate the world of fiction. It seems therefore that Henry James was not merely recording unwritten history, but in some occult way creating it too. The parallels and resemblances validate his claim to verisimilitude. They confirm that there is a truth in art which is truth to life. But they suggest a wider inference as well. Life holds a mirror up to art, because there are fictions that, in reordering the past, create self-fulfilling prophecy. By commanding the imagination they take possession of life, and revert from stories to myths.

# · PART TWO ·

## *History Rewritten*

# · IV ·

## *A Window on the Past*

THERE was a Venice on whose banner was written 'Privacy', and there was another on whose banner was written 'Truth'. The Venice of Henry James was the repository of strange secrets and unwritten history: a territory where faces were masked, where papers were destroyed, and where avenues forever dark and labyrinthine impeded access to the dead. The Venice of Leopold Ranke was altogether different. It was an experience of discovery and elucidation; a window on the hidden landscape of the past. Here secrets came to light, and the seven seals of apocalypse were broken one by one. Ranke, the professor from Berlin who by the 1840s had become the most famous historian of his age, found in Venice the chief means by which he could achieve his life's mission. This was to rewrite the history of modern Europe 'as it had actually been' ('wie es eigentlich gewesen war'). In Ranke's Venice the historian was reborn. He entered the city a discredited judge and fallible prophet; and he emerged from it a revered purveyor of objective truth. He had removed himself from historiography; and he had removed historiography from the flux of history. His gift to the world was, in the estimation of the Victorian historian James Anthony Froude, 'the thing itself, which will have value for all time'.[1] Ranke discovered that from Venice he could see more, and could see further, than he could see from anywhere else; and when he made known what he had seen, Europe looked in gratitude to Venice as the preserver of its memories and the mirror that revealed it to itself.

Ranke introduced a new idol into the temple of European scholarship. That idol was the archive. He taught his generation to mistrust the old historiography, which was based on the materials of libraries—books, chronicles, memoirs, pamphlets, and propaganda—and to demand instead a historiography based on official letters and reports. Only the records of high politics and international diplomacy were unimpeachable evidence of what had happened. It was in state archives that falsifications were unravelled, the passions and partialities of the moment transcended, and the occluded significance of events disclosed. Archives, said Ranke, could reveal 'the true relations of things' and make possible 'a universal history of objective value'. They had mystical properties that revitalized the past. 'Let no-one pity a man who devotes himself to studies apparently so dry,' he adjured his readers. 'It is true that the companions of his solitary hours are but lifeless paper; but they are the remnants of the life of past ages, which gradually assume form and substance to the eye occupied in the study of them.'[2] Other historians were converted by his evangelism, and Victorian professors talking about official documents sounded like Victorian bishops talking about scripture. 'The passage from histories to documents,' declaimed Regius Professor Lord Acton, 'from that which is public to that which is secret, is also the transition from complacent and conventional narrative to the disclosure of guilt and shame.' To enter an archive was to enter the daylight, where error was clearly distinguishable from 'entrusted truth'. 'In proceeding from book to manuscript, and from library to archive', Acton told students at Cambridge in the 1890s, 'we exchange doubt for certainty and become our own masters.' Here were the means to confute 'ill-informed and designing writers, and authorities that have worked for ages to build up a vast tradition of conventional mendacity'. The scholar who endured the pains and penalties of archival research acquired a sort of papal infallibility. 'There is no other way to compel assent or to crush interest and prejudice.'[3] Horatio Brown, likewise, looked to state papers for a revelation of truth and life. 'In an archive', he explained, 'we are as near to the life of the past as it is possible for us to get . . . and our material nearness to the past has a very genuine effect upon the imagination. The naked truth, detailed as it is for no literary purpose . . . produces an effect superior to any that could be obtained by the most skilful master of belles-lettres.'[4]

Venice was renowned for the quality and the quantity of its archival collections. 'Among the archives of Europe,' wrote Horatio in 1886, 'none is superior, in historical value and richness of minutiae, to the archives of the Venetian Republic preserved now in the convent of the Frari at Venice.'[5] These were the legacy of a highly sophisticated and extensive bureaucracy. Venetian government had been government by the written word. Rawdon Brown described how 'from a very remote period the whole course of government had a tendency to multiply official documents.' The Venetian official, unlike the favourites of despots, had been obliged 'to report in writing every measure he took, every piece of intelligence he received'. Consequently, the Republic had bequeathed to posterity a vast amount of documentary material, most of which had survived intact the upheavals of the Revolutionary and Napoleonic periods. Napoleon had confiscated 2,000 files of foreign-policy documents and taken them to Paris; and then in 1798, when Venice was given to Austria, another forty-four cases of documents had been transferred to Vienna. But there were partial retrocessions in 1805 (when Venice again became French) and in 1815 (when it again became Austrian); and in 1817, after all the city's archival deposits had been moved into the redundant convent of the Franciscans, adjacent to the church of the Frari, the Austrian authorities began to arrange and classify the huge accumulation. The records of the Republic, together with the muniments of the various convents, *scuole*, and other corporations, filled 298 large rooms and galleries, ranged in double rows along more than $3\frac{1}{4}$ miles (6 kilometres) of shelving. The total number of individual volumes and bundles was computed at something like 12 million.[6]

The superlative value of these collections lay in the fact that they were much more than just a record of Venetian affairs. The Frari was, in the words of Rawdon Brown, 'a vast magazine of universal history'. This was because the Republic had matched the sophistication of its bureaucracy with the percipience of its diplomacy. It had had eyes and ears everywhere, and the making of modern European history had been witnessed by its perspicacious envoys. During nearly four centuries the Venetian ambassadors had been famous for their skills of observation, inquiry, understanding, and reporting; above all, of reporting. Every ambassador, as well as sending regular dispatches (*dispacci*) to the doge, had been required to deliver a general report (*relazione*) to the Senate on his return. In the surviving records of the Secret Chancery there were over a thousand volumes of *dispacci* from France, England, Milan, Germany, and Constantinople alone, as well as sixteen volumes of *relazioni* on those states. The Secret Chancery records were incomplete, because a fire in the Ducal Palace in 1576 had destroyed the *dispacci* and *relazioni* dating from before the middle of the sixteenth century. Furthermore, there were gaps in the surviving series of *relazioni*. However, the lacunae could in many cases be filled. Ambassadors had often made duplicates of their *relazioni*, and occasionally of their *dispacci*, for private use; and in the difficult times following the collapse of the Republic many of these copies had been sold by their descendants to libraries in Italy and abroad. Duplicates of Venetian documents were to be found in Rome, Paris, Naples, Siena, Florence, Turin, Madrid, London, Berlin, Gotha, and Oxford, as well as in the Marciana Library in Venice. So the searcher for archival material had a second resource when the supply failed in the archives themselves. Taken together, the Venetian *dispacci* and *relazioni* formed a unique quarry of information and analysis. An ambassador's dispatches made up a diary of events, negotiations, and gossip;

while his 'relation' was a widely ranging survey. In the words of the nineteenth-century historian Charles Yriarte:

elle résume, elle conclut, elle expose les tendances de la politique, dit le génie de la nation, le caractère et les facultés des princes . . . Son programme est vaste, tout est de son domaine: le royaume, ses lois, ses mœurs, ses resources économiques, la noblesse, le clergé, le peuple, le conseil, les rapports avec les puissances.*[7]

Habsburg administration was jealously vigilant and elaborately formal. While Venice remained Austrian therefore, access to the records was closely controlled. Ranke had to solicit influential references and endure a personal interview with Prince Metternich before he was granted permission. The French historian Armand Baschet had to apply for ministerial sanction through the French ambassador to Vienna.[8] But, as Baschet discovered, in Venice the scholar was sure of sympathy and support from local officials, and this meant that entry to the archives was always much easier than it was in Vienna, where Ranke was refused permission to read eighteenth-century dispatches as late as 1863. In Venice the general principle of secrecy had been dropped, and the Italians were following where the Austrians had led when they allowed in foreign students virtually without restriction. Under the directorship of Bartolomeo Cecchetti (1876–89) the Frari became a rendezvous of international scholarship. Only the Spanish archives at Simancas, opened to foreigners in the 1840s, rivalled those of Venice in scope and detail; but elation and expectation grew as government after government followed the Venetian example. Doors were unlocked, knowledge was recovered, and Europe experienced a new Renaissance. 'Nearly all the evidence that will ever appear is accessible now,' Acton told contributors to the *Cambridge Modern History* in 1898. 'We approach the final stage in the conditions of historical learning. The long conspiracy against the knowledge of truth has been practically abandoned.'[9] History clamoured to be rewritten as scientific verification replaced imagination and rhetoric in the historian's repertoire. 'I see the time approach', Ranke had said in 1845, 'in which we shall no longer have to found modern history on the reports even of contemporary historians . . . still less, on works yet more remote from the source; but on the narratives of eye-witnesses, and genuine and original documents.'[10] Forty years later Professor Sir John Seeley was declaring that the prediction had been fulfilled. 'It seems not to be generally known', he told Cambridge students in 1883,

* it summarizes, it concludes; it explains the drift of the politics of the country, and talks of its national character, and of the disposition and faculties of its princes. Its topic is vast, everything comes within its scope: the kingdom, its laws, its economic resources, the nobility, the clergy, the people, the council, the relations with foreign powers.

how much the study of history has been transformed of late years. Those charges of untrustworthiness, of pompous and hollow conventionality, which are vulgarly made against history, used to be well-grounded once, but are in the main groundless now. History has been in great part rewritten; in great part it is now true.[11]

It was generally agreed that the new Renaissance had begun with Ranke's discovery of the Venetian records. Ranke described himself as 'the Columbus of the *relazioni*', as the first exploiter of these 'unused, almost unknown' materials, and younger historians were happy to acknowledge his role as pioneer and leader. Ranke, wrote Baschet, 'est . . . le promoteur sans rival de l'importance et de la valeur des textes italiens . . . il est le révélateur du mérite et de la valeur des *relazioni*; il a porté sur elles l'attention des chercheurs; il les a sorties des ténèbres de l'oubli . . .'* Acton described how Ranke discovered 'a new vein' in the Venetian material, and thereby became 'the real originator of the heroic study of records . . . the most prompt and fortunate of European path-finders'. The Venetian manuscripts always claimed Ranke's special reverence and affection. He described them as flowers apparently withered, yet still full of scent and honey. The *relazioni* and *dispacci* were of unique value to him because they both assisted his search for truth and satisfied his yearning for a synthesis of the particular and the universal. Here it seemed were both accurate observation and comprehensive vision. These disinterested, judicious ambassadors, commenting on 'the most important states in the world', enabled the historian to see beyond those 'particular views of parties or of private individuals' that had so distorted earlier synoptic accounts of the modern period—Sismondi's, most notably. Their revelations enabled him to expound and to explain that shift of power and civilization from the Mediterranean to the North that was the most remarkable feature of recent European history.[12]

Ranke first encountered *relazioni* in the winter of 1824–5, when he was a fledgling professor (not yet 30) enjoying the success of his first book. He had heard about a collection of Italian documents in the royal library at Berlin, and when he investigated he found forty-eight folio volumes containing sixteenth- and seventeenth-century manuscripts, of which a large number were transcripts of Venetian *relazioni*. They were reports on the Ottoman Empire and the court of Spain, on Piedmont, Tuscany, Urbino, Naples, and on the papal court in Rome, written by ambassadors who were among 'the most able and experienced citizens' of the Venetian Republic, and who had acquired ample stores of that insiders' knowledge 'which is

* Ranke is the unrivalled propagator of the importance and value of the Italian texts . . . he is the revealer of the merit and value of the *relazioni*; he has drawn them to the attention of researchers; he has brought them out of the shadows of oblivion.

hidden from the crowd and which dies with themselves'. He found a further four volumes of *relazioni* in the ducal library at Gotha, and told his publisher of his intention to write a history of modern Europe based on these 'unknown, indubitable, and most interesting reports'. The first results of his new researches were published between 1827 and 1836 under the general title *Fürsten und Völker von Sud-Europa im Sechzehnten und Siebzehnten Jahrhundert*,* but only the first part of this survey—two volumes on the Ottoman and Spanish Empires—was written from material available in Germany. Ranke said that after working in the German libraries, he was more acutely conscious of what was missing than of what had been found. 'In the midst of wealth we are sensible of our poverty. As a whole there is much; but when we look to particulars, great wants are apparent.' His information on the Italian states was especially patchy, and he felt like a traveller who had begun to redraw the map of a dimly known territory 'and has no more earnest wish than to return and make his inspection complete'. The full documentary evidence was available only abroad, so his return to the half-discovered country became a journey through space as well as through time. At the end of 1827 he left Germany on an archival tour that was to last three and a half years and result in the publication of *Die Römischen Päpste, ihre Kirche und ihre Staadt im Sechzehnten und Siebzehnten Jahrhundert*.† This was the first of his books to be translated into English (1840) and the one that established the international fame both of its author and of the Venetian records. Ranke went first to Vienna. Here he found a large collection of Venetian manuscripts, which included many copies, and some originals, of *dispacci* and *relazioni*, as well as chronicles and diaries. Then in October 1828 he arrived in Venice itself and discovered the treasures of the state archives. He spent nine months reading and transcribing in the Frari, before moving on to the libraries of Rome and Florence. He spent a further three months in Venice on his way home, and returned to Berlin in the spring of 1831 gorged with the archival nutriment that he would spend the rest of his long life digesting and secreting. After the three volumes on the Popes came six volumes on Reformation Germany (*Deutsche Geschichte im Zeitalter der Reformation*), four volumes on French history (*Französische Geschichte, vornehmlich im Sechzehnten und siebzehnten Jahrhundert*), and nine volumes on English history (*Englische Geschichte, vornehmlich im Siebzehnten Jahrhundert*)—all of them indebted in greater or lesser degree to the sage Venetian witnesses. And the posture that Ranke adopted from these ambassadors—

* *Princes and Peoples of Southern Europe in the Sixteenth and Seventeenth Centuries.*

† *The Popes of Rome, their Church and their State in the Sixteenth and Seventeenth Centuries.*

that of the detached observer, the clinical recorder of facts, of the thing itself—he passed on not only to other historians, but to novelists and artists as well. In the 1890s the French historian Gabriel Monod saw in retrospect how scientific realism had become 'le principe organique de la vie intellectuelle en France'. He detected it in the pictures of Meissonier, Millet, and Bastien Lepage, and in the fiction of Flaubert, Maupassant, Zola, and Bourget, as well as in the historiography of Renan and Taine.[13]

After Ranke had shown the way, a generation of students looked to Venice for enlightenment. The Italian scholar Eugenio Alberi used *relazioni* for his life of Catharine de' Medici (1838). In the early 1840s the records of the old Secret Chancery were opened for the French historian Louis de Mas Latrie, who was writing a history of Cyprus under the Lusignan dynasty. Ten years later Ignaz von Döllinger, eminent historian of the Church, arrived from Munich with the youthful Acton, who was at this time his brilliant pupil. In 1864 Acton returned alone to investigate the temporal power of the papacy and the doctrine of Papal Infallibility, and worked, according to Rawdon Brown, 'at the rate of seven hours per diem in the library besides his occupations at home'. Brown was deeply impressed by Acton's erudition, industry, memory, and regard for truth. 'During my thirty years' stay here', he told Layard,

> I have never met with anybody who more thoroughly understood the value of Venetian historical remains or availed himself of them with better method, to say nothing of the assiduity with which he works from morning to night. In my opinion he bids fair to become the best modern historian (from 1500 to 1800) of this century.

In the early 1860s the American historian John Lothrop Motley also was in Venice. He asked William Howells to obtain copies of *relazioni* from the Frari, and Howells later built a novel round the incident. In *A Fearful Responsibility* Professor Owen Elmore comes to Venice to pursue his research away from the turmoil of the Civil War, and applies for leave to have transcripts made in the archives. 'The permission was negotiated by the American Consul (then a young painter of the name of Ferris), who reported a mechanical facility on the part of the authorities.'[14] The celebrity of the Venetian records was such that they began to be printed, and historians who could not go to the archives enjoyed the privilege of having the archives brought to them. An uncoordinated but dedicated effort of international scholarship made the *relazioni* and *dispacci* easily obtainable. Italian, French, Austrian, and English editors compared, deciphered, translated, and calendared the surviving texts, and both private and public money financed publication. In 1833 the French historian and statesman François Guizot, then minister in charge of national education, explained to King Louis Philippe how in recent years 'l'étude des sources historiques [avait] repris une activité nouvelle'.* He petitioned, successfully, for funds for the publication of a series of *Documents inédits relatifs à l'histoire de France*, and among the scholars and men of letters commissioned to collect and prepare material was Nicolò Tommaseo, a Venetian political exile living in Paris. Tommaseo produced two volumes of translations of *relazioni* referring to sixteenth-century France; but he worked from copies in the Bibliothèque Royale, and his texts were criticized as incomplete and garbled. In order to obtain accurate transcripts the historian Armand Baschet was sent to Venice by the French government in 1852. Here he spent five years, collecting enough material on sixteenth- and seventeenth-century France to fill, he claimed, eight volumes—though only two were in fact published.[15] Meanwhile a group of learned and literary men in Florence, deeply impressed by the use that Ranke had made of *relazioni* in his history of the popes, resolved to put the whole of them into print. The marchese Gino Capponi, who had many transcripts in his private library, rallied subscribers and supporters, and Eugenio Alberi was appointed editor.[16] Alberi scoured libraries in Venice, Milan, Turin, Rome, Vienna, Berlin, Paris, and Gotha, and published fourteen volumes of sixteenth-century *relazioni* between 1839 and 1863. A further ten volumes, covering the seventeenth century and reproducing original texts from the Frari, were published in Venice between 1856 and 1878 under the editorship of Nicolò Barozzi and Gugliemo Berchet. One of Acton's activities in the 1860s was assisting Tommaso Gar, director of the Venetian Archives, in editing the *dispacci* of Antonio Giustinian, ambassador to the Holy See from 1502 until 1505. Four volumes of *relazioni* on Germany and Austria appeared in Vienna, in German translation, in the 1860s, and one of these contained the only eighteenth-century reports to appear in print before the 1940s.

The first English translations of ambassadorial documents appeared in 1854, when Rawdon Brown published *Four Years at the Court of Henry VIII*. This was two volumes of selections from the dispatches of the ambassador Sebastian Giustinian. The originals had been lost, but Brown discovered copies of 226 letters among the Contarini manuscripts in the Marciana Library. The book revealed the interest of the Frari archives for students of British history, and in 1862 the London government decided to sponsor a calendar of all Venetian state papers relating to British affairs. For

* the study of historical sources [had] taken on a new activity.

the next twenty years Brown was employed, at £200 a year, in working steadily through the diplomatic correspondence of the Secret Chancery, 'calendaring every document in which the name of any one of the three kingdoms appears, or any of their subjects is mentioned'.[17] He edited six volumes, three of them double. The seventh, which brought the calendar to 1580, was finished after his death by his friend and executor George Cavendish Bentinck.[18] Layard, who was under-secretary for foreign affairs from 1861 till 1866, was clearly behind the government's decision to sponsor the calendar. He corresponded with Rawdon Brown on technical matters relating to the manuscripts, and in 1889 used his influence to get Horatio Brown appointed editor in place of Bentinck. 'Both my mother and I', Horatio wrote to Layard in March 1889, 'are anxious to see how the question of the appointment will end. Mr Bentinck has stuck to the post so long that until he is absolutely and definitely out of it I hardly like to think of the matter as concluded. . . . I wish to thank you now, Sir Henry, for your constant help and unfailing kindness to me.'[19] In June 1894 he told Layard that he would shortly be sending him a copy of his first volume. 'I am almost in hopes', he added ingratiatingly, 'that the Constantinople despatches may interest and even amuse you.'[20] From 1889 until 1905 Horatio reserved the labours of his own writing and the pleasures of Venetian life for the afternoons and evenings. His mornings, from soon after nine o'clock until just before one, he reserved for the work of deciphering, transcribing, and epitomizing in the Frari. He edited five volumes of the *State Papers Venetian*, as they were known, and brought them down to 1613. In all, thirty-eight volumes of the calendar were published, and when it was abandoned, in 1947, it covered the period from 1202 until 1675.

Froude prognosticated for Ranke and his school an enduring historiography and a place among the immortals. Yet when the reckoning was made the promise had not been redeemed. In the years immediately before his death in 1886, Ranke seemed like a candidate for apotheosis. The octogenarian scholar, blind and unable to hold a pen, dictating the great *Weltgeschichte* that was to be the summing-up of a lifetime's learning, had acquired Homeric grandeur. But his books did not survive. Most of them were dead before he was, and if they still had interest in the next century this was only because they had themselves become historical phenomena. As early as 1867 Acton was composing the obituary of his most famous work. 'The *Popes*', he wrote, 'cannot be a classic for this age as it was for the last. Time makes its defects more glaring.' As Acton pursued his own search among historical evidence for truth and life, so the historiography of Ranke seemed to contain less and less of both. 'The cup is not drained,' he complained in 1886; 'part of the story is left untold, and the world is much better and very much worse than he chooses to say.' By 1895 he was telling students at Cambridge: 'I should not even venture to claim for Ranke . . . that there is one of his seventy volumes that has not been overtaken and in part surpassed.' And opinion bowed before Acton's immense authority. 'Even the most loyal disciples', wrote G. P. Gooch of Ranke in 1913, 'now admit that there were spots on the sun.'[21]

The great man was to some extent a victim of his own success. To the next generation Ranke was not Rankean enough. His reputation was impaired by the archival fetish he had himself created. As historians gained access to more and more state records, and measured the range of his research against the vastness of the documentary legacy, the foundations of his reconstructions began to look perilously meagre. To students who wallowed in archival gluttony, the Venetian *relazioni* were not so satisfying. 'They do not relate current events', complained Acton, 'or grasp history in motion.' Horatio Brown agreed. They were, he pointed out, 'drawn up on more or less conventional lines; the headings . . . were indicated by the Government, and were invariable; and further, the home-coming ambassador handed his report to his successor, who frequently used it as a basis in drawing up his own. The result is, that . . . the *Relazioni* are apt to repeat themselves.' The *dispacci* were an essential complement and corrective; and Ranke was blamed for not making fuller use of them—especially since, in the words of Acton, '[they] were just as easy to consult as the final *Relazioni*'. However, even in their entirety the Venetian testimonies were flawed. 'The Venetians were on the governing side', explained Acton, 'and viewed the state from the point of view of the trouble and difficulty of governing.' What looked like impartiality was in fact official indifference to morals and ethics. Like certain modern statesmen, they saw 'nothing in politics but higher expediency and no large principles'. Because Ranke saw the world through their eyes, he never attained to that 'true impartiality' that 'judges resolutely'. The master began to acquire the reputation of a skimmer, a dilettante, an eker-out of trifles. Flaubert once commented that writing history was like drinking an ocean and pissing a cupful. Ranke it seemed was doing the opposite. As Acton put it, in language more decorous than Flaubert's, 'a little produce went a long way'. The Czech historian Anton Gindely went so far as to call Ranke a charlatan. 'He even resorts to deception,' he wrote, 'wishing to make his readers believe that he has worked through the archives . . . His citations are more crumbs stuck together in a chance fashion to produce the appearance of being the results of systematic study.'[22]

Ranke was also a casualty of fashion. If the chaos of history is unbearable to contemplate, hardly less so is the coherence imposed on it by historians of the previous generation. The works that Ranke presented as synoptic histories were reckoned by younger scholars to be narrow and selective. 'There is a tendency', complained Gooch, 'to survey events too much from the council chamber.' *Staatengeschichte* was out of date; the historiography now favoured was *Kulturgeschichte* of the sort written by Rhiel, Freytag, and, above all, Burckhardt. 'History has acquired a much ampler and more comprehensive meaning,' explained J. B. Bury, who succeeded Acton as regius professor at Cambridge in 1903. 'Every form of social life and every manifestation of intellectual development should be set forth in its relation to the rest.' Karl Lamprecht, professor at Leipzig, demanded a 'psychological' historiography, in which 'description alone' would be superseded by 'an intelligent comprehension'. It was now a question, he said, 'of following up the complex phenomena of the socio-psychic life, the working out of the so-called national soul in its elementary parts'.[23]

Acton's indictment of Ranke drew on something deeper than fad or pedantry. It was the effect of a cause that undermined the faith by which both he and Ranke had lived. Although a Catholic, Acton had a Protestant's obsession with truth. 'The full exposition of truth', he wrote in 1863, 'is the great object for which the existence of mankind is prolonged on earth.' Furthermore, he assented to the Rankean dogma that truth was to be found in archives. Nothing could sound more ringingly dogmatic than Acton saying: 'In proceeding from . . . library to archive we exchange doubt for certainty and become our own masters.' But Acton's experience was the reverse of Ranke's. Few eminent careers have shown so little evidence of certainty. Uncertainty drove him to find out what countless others had said before he spoke himself. Uncertainty cramped his utterance into embryonic densities. Uncertainty enmeshed him in endless preliminaries. The Rankean music-drama unfolded immense lengths; but there was Actonian overture after Actonian overture and the curtain never rose. Ranke thought in volumes. Acton thought in apophthegms, and could not sustain more eloquence than would fill a lecture or an article. Fluency was curtailed, expatiation checked, beginnings postponed by the crippling conviction that he did not know enough. He felt disqualified partly because there was now so much to know. 'Modern history', he said, 'is buried under myriads of documents yet unseen, and . . . the composition of books fit to live must be preceded by the underground labour of another generation.' The sources were now so copious that there was 'more fear of drowning than of drought'.

However, his difficulties were related as much to the quality as to the quantity of his evidence. The more Acton read and researched, the more aware he became of a seventh seal intact. He was haunted by the idea of some final secret, some elusive, residual truth that retreated as the historian advanced, leaving him perpetually tantalized amidst 'conventional mendacity'. 'After all that', he wrote of modern archival research on the French Revolution, 'we are still ignorant of the most essential things.'[24]

He drew a distinction between history and biography. The province of the one was public, of the other private life. The historian's concern was with crime, not vice. It was not his function to judge sins 'which disturb only the relationship of man with God'. 'Looking at history', he told Mary Gladstone in 1882, 'taking societies, and not individuals, we cannot deal with things seen by God alone. . . . The scale of vice and virtue is not that of private life. . . . The test and measure of good and evil is not that of the spiritual biographer.' Yet Acton demanded admission to the confessional. Although he might forbear to judge, the historian must nevertheless know and understand the inner workings of the soul. He must, like the novelist, be omniscient. 'My life', he wrote to Mary Gladstone, 'is spent in endless striving to make out the inner point of view, the *raison d'être*, the secret of fascination for powerful minds of systems of religion and philosophy, and of politics . . .' He conceded that he was divided from George Eliot 'by the widest of all political and religious differences'; nevertheless he reverenced her for her profound knowledge of her characters, since that was the condition of true impartiality. She was capable

> not only of reading the diverse hearts of men, but creeping into their skin, watching the world through their eyes, feeling their latent background of conviction, discerning theory and habit, influences of thought and knowledge, of life and of descent; and having obtained this experience, of recovering her independence, stripping off the borrowed shell, and exposing scientifically and indifferently the soul.

In his own striving for omniscience Acton looked beyond archives and official documents. He devoured memoirs as they were published. 'The reality of history', he wrote in 1891, 'is so unlike the report that we continue . . . to look for revelations as often as an important personage leaves us his reminiscences.' He was avid for unpublished personal papers, believing that 'no public character has ever stood the revelation of private utterances and correspondence'. 'The great addition' to official sources was 'the unpremeditated revelation of correspondence'. And from this appetite for information developed the mentality of the eavesdropper, of the spy at the keyhole. 'In modern times', he wrote in 1890,

1. The Bacino di San Marco in the 1880s.

2. (*a*) The Zattere in the 1880s. The façade of the Gesuati just intrudes on the right.

2. (*b*) The Grand Canal in the 1880s, viewed from outside the Accademia.

3. (*a*) The original Accademia Bridge.

3. (*b*) The Campo di Santa Marta before the demolitions of 1883. This became the site of the cotton factory.

4. A courtyard in the area of San Pantaleone, late nineteenth century.

5. The façade of the Ca'd'Oro after late nineteenth-century restoration.

6. The Palazzo Dario in the late nineteenth century.

7. The Palazzo Rezzonico at about the time of its acquisition by Pen Browning.

8. Ca' Cappello, with the Layards' gondola at the door.

9. (*a*) Pen Browning.

9. (*b*) Fannie Browning.

10. John Addington Symonds and his daughter Margaret (Madge).

11. Horatio Brown in middle age and as a young man.

12. (*a*) The southern angle of Sansovino's Libreria di San Marco.

12. (*b*) The West of England and South Wales Bank, Bristol, by William Gingell and T. R. Lysaght. A Victorian paraphrase of Sansovino's Library.

13. (*a*) Campanile and Basilica di San Marco in the early 1860s. Meduna's restoration is in progress at the south-west angle.

13. (*b*) The south-west angle of San Marco in the 1880s, showing the reinstated Zeno chapel.

14. (*a*) The southern façade of the Ducal Palace in the 1860s. The furthest five arches, walled up after the fire of 1577, were opened during restoration 1876–1889.

14. (*b*) Sansovino's Loggetta, at the base of the Campanile, in the 1880s.

15. (*a*) The ruins of the Loggetta, after the collapse of the Campanile, July 1902.

15. (*b*) The Piazza with the debris of the Campanile, July 1902.

16. Venice preserved. The official inauguration of the rebuilt Campanile, April 1912.

since Petrarca, there are at least two thousand actors on the public stage whom we see by the revelation of private correspondence. Besides letters that were meant to be burnt, there are a man's secret diaries, his autobiography and table-talk, the recollections of his friends, the self-betraying notes in the margins of books, the report of his trial if he is a culprit, the evidence for beatification if he is a saint.

When he was planning the *Cambridge Modern History* he anticipated unprecedented disclosures in the sections dealing with the recent past. 'The last volumes', he promised the syndics of Cambridge University Press, 'will be concerned with secrets that cannot be learnt from books, but from men.' He advised the contributors: 'Certain privately printed memoirs may not be absolutely inaccessible, and there are elderly men about town gorged with esoteric knowledge.'[25]

He has moved from the Venice of Ranke to the Venice of Henry James. Acton the regius professor begins to sound like a 'publishing scoundrel', pleading 'the rectification of history' in his eagerness to expose the face behind the mask, the person behind the performance. When his contemporary James Bryce sat down to write a biographical memoir of Acton, Jamesian images and metaphors came irresistibly to mind:

He . . . held that the explanation of most of what has passed in the light is to be found in what has passed in the dark. He was always hunting for the key to secret chambers, preferring to believe that the grand staircase is only for show and meant to impose upon the multitude, while the real action goes on in hidden passages behind. No-one knew so much of the gossip of the past; no-one was more intensely curious about the gossip of the present, though in his hands it ceased to be gossip and became unwritten history.[26]

And perception of the same affinities seems to have influenced Mary Gladstone. She borrowed the title of one of James's stories, *The Madonna of the Future*, for Acton's grandly conceived and forever delayed history of liberty.

The experience of the Jamesian protagonist, in life as in fiction, was one of frustration and paralysis. His route was barred by jealous guardians of reputations and all the contrivances with which privacy baffles truth. He encountered what Sainte-Beuve had called 'la réserve imposée aux témoins contemporains, et la dignité de l'histoire'.[27] Hopes of elucidation from published reminiscences consequently led only to 'much disappointment'. Talleyrand's memoirs, finally issued in 1891, provided yet another instance of anticipation betrayed. 'The famous book which has been so eagerly expected and so long withheld', wrote Acton in his review, 'will not satisfy those who, like the first Queen of Prussia, demand to know *le pourquoy du pourquoy*. . . . The most experienced and sagacious of men . . . betrays few secrets and prepares no surprises.'[28] Because Acton was naïve, his life was full of irony, and most exquisitely ironical of all was his association with Horatio Brown. He recruited Brown as a contributor to the *Cambridge Modern History* and counted him an accomplice in the task of unearthing the secrets of the past. Yet it was this fellow historian, this partner in the building of the great monument to truth, who was in his own life one of the most culpable offenders against its spirit. Few elderly men about town were more gorged with esoteric knowledge; and none, probably, revealed so little and destroyed so much. Of this treachery Acton never learnt; but he was always aware of the insufficiency of the record, and in his heart of hearts he seems to have believed that belief in the resurrection of the past was the vanity of vanities. As Gooch so acutely observed, there is no better description of Acton than the description that Acton himself supplied of his old teacher Döllinger: 'His way was strewn with promise unperformed, and abandoned from want of concentration. He would not write with imperfect materials, and to him the materials were always imperfect.'[29]

## · V ·

# *Retreat from Legend*

THE collapse of Venice in 1797 and its transformation into a puppet democracy, a satellite of France, were incidents almost without consequence among international events. At the time of its abdication the old Venetian Republic was impotent and ignored, a political anachronism of little interest and less influence. But when Napoleon delivered the infant Venetian democracy to Austria, later the same year, a new preoccupation was added to the concerns of the modern mind. The treaty of Campo Formio formally ended the independence of Venice; and as soon as the state was dead, its history came to life. An obscure chronicle turned into a momentous epic. Its ignominious extinction reminded the world that this nonentity, this lingering vestige of a defunct supremacy, had been the oldest and most extraordinary polity in Europe. It had been a city-state among territorial sovereignties; a mercantile republic among feudal monarchies; a token of continuity amid all the vicissitudes of history from the fall of the Roman Empire to the outbreak of the French Revolution. The contrast between the magnitude of its former importance and the pathos of its present condition called for lofty, Gibbonian periods. 'Resembling no other political edifice of ancient or modern creation,' wrote a reviewer called George Procter in the *Quarterly* in 1825, 'it stood in the days of its grandeur an object of silent astonishment to the nations; and its ruins remain on the broad waste of time to attract the eye of philosophical research or to kindle the splendid associations of romance.' And Venice added to the necrology of extinct empires a story of decline and fall whose accomplishment was all the more suggestive in that it closed an existence of a thousand years. The completed saga of the Republic was not only symmetrical, it was millenarian; not just aesthetically satisfying, but deeply portentous. So Venice became more interesting when it ceased to be topical. It attracted poets, novelists, and librettists, as well as narrative and scientific historians; and they all produced a literature of 'romance' and 'philosophical research' in which the Venetian past, refracted by present preoccupations, became an index of national concerns and modern mentalities.

The thirst for explanation that followed this latest instance of the mutability of earthly dominion was initially satisfied for Anglo-Saxon inquirers by a theory of guilt and divine retribution. The record seemed to confirm the opinion that had been gaining currency in the later seventeenth and during the eighteenth century. Venice had been tyrannical and corrupt, not glorious and exemplary in the way that its apologists had made it out to be. Rousseau's verdict on the Venetian junta in *Du Contrat social* ('un tribunal de sang, horrible également aux patriciens et au peuple'*) was endorsed, and it became a commonplace of educated opinion that the fall of Venice had been a punishment for heinous crime. The Republic came to figure in popular British and American literature as sinister, cruel, and inscrutable. No other government could match the wickedness with which it had, in the words of the historian Henry Hallam, 'set aside the rights of man and the laws of God'. Even Byron sounded like an Old Testament prophet when writing of its fall. In his Venetian verse-dramas—*The Two Foscari* and *Marino Faliero*—he depicted a godless world of high politics in which the individual was crushed by pitiless *raison d'état*; and *Marino Faliero* ends with the condemned hero predicting the chastisement and humiliation that await this 'Gehenna of the waters', this 'sea-Sodom', in the unfolding of history:

> the hours
> Are silently engendering of the day
> When she, who built 'gainst Attila a bulwark
> Shall yield, and bloodlessly and basely yield,
> Unto a bastard Attila.

James Fenimore Cooper's popular historical novel *The Bravo* (1831) encapsulates a vision of totalitarian power that anticipates the nightmare of Kafka and Orwell. It portrays Venice as a police-state where anonymous, faceless authority, employing means both horrible and terrible, punishes the noblest human instincts and panders to the most sordid in order to extirpate sedition. 'The reader will at once see', commented Cooper,

> that the very reason why the despotism of the self-styled Republic was tolerable to its own citizens was but another cause of its eventual destruction. . . . Each lived for himself, while the State of Venice held its vicious sway, corrupting alike the ruler and the ruled, by its mockery of those sacred principles which are alone founded in truth and natural justice.

* a tribunal of blood, horrible to patricians and people alike.

Edward Smedley's *Sketches from Venetian History*, published in 1831 and later reissued as part of Murray's Family Library, was for almost half a century the only synoptic narrative in English. By selecting 'the most striking incidents' in the Venetian record, and 'connecting them with each other by a brief and rapid survey of minor events', Smedley constructed a catalogue of malefaction that culminated in 'the hour of visitation'. 'Never yet', declaimed this Anglican clergyman,

> did the Principle of Ill establish so free a traffic for the interchange of crime, so unrestricted a mart in which mankind might barter their iniquity; never was the committal of certain and irremediable evils fully authorized for the chance of questionable and ambiguous good; never was every generous emotion of moral instinct, every accredited maxim of social duty, so debased and subjected to the baneful yoke of an assumed Political expediency.

With such powerful denunciations ringing in their ears, early Victorian observers of prostrate Venice could draw only one conclusion. 'Let a man see all this,' commanded Frederick Faber in 1841, 'and . . . let him see even upon the blighted greatness of these Adriatic lagoons the righteousness of God.' John Ruskin saw it vividly. He linked the fall of Venice with the fall of Tyre. 'The evidence which I shall be able to deduce from the arts of Venice', he announced, 'will be both frequent and irrefragable, that the decline of her political prosperity was exactly coincident with that of domestic and individual religion.' Venice was doomed, according to Ruskin, when it ceased to atone for public crimes with private piety.[1]

This Anglo-Saxon response was an evangelical reading of the work of French historians. Bonaparte's treatment of Venice had troubled even his supporters.[2] They looked to history for explanation and excuse, and the evidence they amassed to vindicate Napoleon was used by the British and Americans to prove the righteousness of God.

Léonard Sismonde de Sismondi's *Histoire des républiques italiennes du moyen âge*, published in sixteen volumes between 1809 and 1818, presented the history of Venice as the tragedy of a people who, once strong because they were free, became weak and craven because they were enslaved. It was the story of democracy supplanted by aristocracy, and of aristocracy supplanted in its turn by the abstract collectivity of the state. Since the early fourteenth century Venice had been ruled by a despotic commission that violated individual rights of life, liberty, and property in the interests of the Republic. This was the dread Council of Ten, which was not only a cause but a symptom of turpitude, since the aristocracy could have voted it out of existence at any time. Only an abject and demoralized nobility, argued Sismondi, would have tolerated for five hundred years its régime of spies, police, torture, and arbitrary and secret executions.[3] By the time he finished his book, Sismondi had become an enthusiastic admirer of Napoleon. Pierre Daru, who amplified and substantiated Sismondi's thesis, had been one of his most indefatigable and trusted (though not so trustworthy) civilian coadjutors, first as *chef de division* at the war ministry, and subsequently as *ministre secrétaire d'état* and *intendant général* of the imperial household. He had literary ambitions and connections—Stendhal was a relative and protégé—and after the restoration of the Bourbons, when he was temporarily banished from Paris and from public life, he occupied his leisure by writing the first complete history of Venice. This was published in eight volumes between 1815 and 1819. Three further editions followed, and Daru was for more than thirty years recognized as the leading authority in his field. Although his book was never translated into English, it became as well known in Britain and America as in France and Italy.[4]

Daru acknowledged the epic in his subject: the story of a handful of fugitives who founded a famous city on the Adriatic mud; sent out fleets to dominate the seas; overturned the mighty empire of the Turks; brought back the riches of the Orient; and arbitrated the fortunes of Italy. 'C'est là, sans doute,' he conceded, 'un développement d'intelligence humaine qui mérite d'être observé.'* But he was insensitive to Venetian civilization and he judged its degradation coldly, as a sordid saga of imperial hubris and moral collapse. Daru interpreted the history of Venice as Montesquieu and Gibbon had interpreted the history of Rome. Civic virtues had been forgotten and liberty lost in an orgy of conquest and luxury; and all that remained as a legacy to humanity were 'quelques découvertes dans les sciences et quelques monuments des arts'. Of these the most notable was the printing press. Any truly great artistic or intellectual achievement had been precluded by repressive government and an all-pervasive mercantile spirit. It was this spirit that explained the Venetian acquisition of empire, since commerce can never stand still: 'Le commerce, cette profession où l'on tente continuellement la fortune, n'est pas une école de modération. Les succès inspirent l'avidité et la jalousie, et celles-ci l'esprit de domination. . . . Cet esprit d'ambition est au fond le même que celui des conquêtes.'† And cupidity had been assisted by ingenious statecraft. Venice had enjoyed political advantages that its rivals could not match. Its government had been stable, efficient, and astute; astute above all in its encourage-

* Here, doubtless, is a development of human intelligence that merits notice.

† Commerce, the calling in which one constantly tempts fortune, is not a school of moderation. Success inspires avidity and jealousy, and these in their turn arouse a spirit of domination. . . . This spirit of ambition is fundamentally the same as that of conquest.

ment of private vice in order to pre-empt political ambition, and in its use of propaganda and spectacle as a means of psychological manipulation. 'Tout ce spectacle de grandeur, de richesse, de joie',[5] explained Daru, 'animait une population active et ingénieuse, et lui inspirait un trop juste orgueil pour qu'elle ne dût pas se croire contente de sa destinée.'*

However, the price that had to be paid for all this éclat and pre-eminence was heavy. Emasculated by wealth and vice, dazzled by illusions of freedom and power, the Venetian patriciate, having usurped the rights of the Venetian people, were in their turn ensnared by the vilest despotism that the world had ever known—a despotism that completed their moral vitiation and made inevitable their ruin. It was Daru's special claim that he exposed for the first time the inner workings of the old Venetian government. The Council of Ten and its nefarious retinue of police, informers, and spies were already familiar. Daru now revealed to an appalled world the true nature of the Inquisition of State, an elusive and mysterious triumvirate that was known to have functioned as a committee of the Ten from the beginning of the sixteenth century. Among the manuscripts in the royal library in Paris he discovered a transcript of the statutes of this institution, which turned out to be one of only five copies, all of them hitherto unknown to historians. This text, which Daru published *in extenso*, revealed that the Inquisition dated from 1454 and confirmed that it had been the quintessence of totalitarian tyranny. Omnipresent, arbitrary, immovable, 'ce tribunal monstrueux' had exercised a jurisdiction that no one escaped—neither the Ten, nor the doge, nor even the inquisitors themselves. It had had eyes and ears everywhere, and there had been neither appeal from its verdicts, nor redress for its errors, nor access to its proceedings. 'Un homme disparaissait; et, si l'on pouvait soupçonner que ce fût par l'ordre de l'Inquisition, ses proches tremblaient de s'en informer.'† In cases of capital punishment, poison and drowning had been among its modes of execution.[6]

Daru contended that the decline and fall of the Republic had been directly and principally attributable to its evil constitution. This was at the root of the moral decadence that had made the Venetians such easy prey to adverse circumstances. Adversity would have been less powerful if they had been less effete. He therefore disputed that the economic decline of Venice was inevitable after the discovery of the Cape route to India and the shifting of the world's trade from the Mediterranean to the Atlantic. 'Qui l'aurait

* All this show of greatness, wealth, and joy encouraged an active and ingenious people, and inspired in them a pride too justified for them not to believe themselves happy with their lot.

† A man disappeared; and if it was suspected that it might have been by order of the Inquisition, his kin trembled to inquire further.

empêchée', he demanded,[7] 'de porter son pavillon dans l'océan, comme les Portugais, les Espagnols, et les Hollandais?'*

Stendhal accepted Daru's vindication of Napoleon and wrote of 'la barbarie froide' of the old Venetian Republic.[8] In Britain the *Quarterly*'s reviewer, George Procter, reproached the historian for chauvinistic bias in his final chapters. 'He betrays the leaven of the revolutionary spirit', he wrote, '. . . and his readers might remain profoundly ignorant of the abandoned treachery and insolent aggression of the French rulers in the final catastrophe of Venetian independence.' Daru had become 'the weak apologist for the crimes of the French revolutionary government'. However, Procter did not dissent from Daru's account of the Venetian constitution, and he read his indictment as an invitation to divine chastisement. Venice had perpetrated 'the foulest system of assassination and tyranny, the most deliberate violation of the laws of God and the obligations of morality that ever assumed the shape of human government'.[9] Among Daru's Italian readers was Ugo Foscolo, the Venetian democrat and poet who took refuge in England after the surrender of Venice to Austria. Foscolo, who claimed personal experience of the working of the old Venetian government, reckoned that Daru's account was more accurate than Casanova's. In fact when the first volumes of Casanova's memoirs appeared, in Germany in the early 1820s, Foscolo dismissed them as more fantasy than fact because they were at variance with Daru in certain essential details.[10]

At this stage Daru's only serious critic was Count Domenico Tiepolo, an erudite and impoverished member of the Venetian aristocracy. He denounced Daru's work as an extended exercise in French propaganda; a diatribe designed to justify Napoleon's aggression and treachery. He filled six notebooks with comments and corrections, arraigning the historian for 'inimicizia concepita contro li Veneziani'. He accused Daru 'di voler malignare li Veneziani', 'd'inveire contro li Veneziani', 'di mettere in mala vista li Veneziani', 'di dare un cattivo colore alla condotta de' Veneziani', 'di fare il peggio ritratto possibile del governo di Venezia'. Anti-aristocratic prejudice disfigured all his judgements.[11] 'Di ogni malo, secondo il Sig. Daru, deve essere cagione l'aristocrazia veneta. . . . Ad oggetto di screditare il governo veneto, fa una odiosissima pittura dell'aristocrazia . . . in generale.'† Tiepolo defended aristocracy in principle (was collective rule by

* Who would have prevented her from flying her flag in the ocean, like the Portuguese, the Spanish, and the Dutch?

† animosity conceived against the Venetians . . . of wishing to malign the Venetians, of inveighing against the Venetians, of giving a dark colouring to the conduct of the Venetians, of painting the worst possible portrait of the government of Venice. . . . According to Signor Daru every evil is to be attributed to the Venetian aristocracy. . . . With the object of discrediting the Venetian government he paints a most hateful picture of aristocracy in general.

the best educated not preferable to the tyranny both of monarchs and of mobs?), and refuted the charges of usurpation, cruelty, and immorality that Daru had levelled against the aristocracy of Venice. There had been no usurpation of popular rights, since Venice had been an aristocracy from the beginning, and Daru's idea of the Council of Ten had been derived from biased and ill-informed accounts ('dalle false ed alterate relazioni di quelli che non conoscevano il vero sistema ed oggetto di questa magistratura'). It showed him 'di non aver la minima conoscenza dell'organizazione del governo veneto, nè antico nè moderno'.* His description of the notorious cells of the Ducal Palace, where prisoners were supposed to have either rotted in the damp (in the *pozzi*) or suffocated in the summer heat (in the *piombi*) was melodramatic nonsense. When the Republic fell the *pozzi* had been out of use for many years, and the amenability of the *piombi* was proved by the fact that they had now been converted into commodious government offices. The accusations of moral corruption and of decadence were so lacking in substance and so inflated with fiction that it would require a dissertation to refute them. This Tiepolo did not attempt. Instead he deployed his polemical skill against the so-called statutes of the Inquisition of State. These, he argued, were a forgery. Daru had been duped. There was no trace of them in the State Archives in Venice, and the transcript quoted was full of incriminating anachronisms. Though dated 1485, they were written in the Venetian dialect, which was not used for official documents until half a century later. They referred to 'Inquisitori di Stato', a title not adopted until the seventeenth century. They classified the *piombi* as a state prison, whereas these were not used as such until 1591. And so on. Tiepolo's scorn was withering. Daru's evidence was worth no more than 'popolari tradizioni che non meritano credenza che dalle vecchiarelle e da' fanciulli'.†

Daru asked to see Tiepolo's manuscript and received a copy of the 228 *osservazioni* in March 1823. He accepted a few of them and amended his text for the third, 1826, edition of his book. The others, including those relating to the statutes, he rejected, and Tiepolo was sent a large notebook full of rejoinders. It was at this point that Tiepolo decided to publish his critique. It came out in two volumes in 1828, and was corroborated by Bianchi Giovini in 1837, when he published an Italian translation of Daru's text. Meanwhile Ranke too had impugned the statutes, in an essay of 1831 entitled *Über die Verschwörung gegen Venedig in Jahr 1618* ('On the Conspiracy against Venice in the Year 1618'). By now Daru was dead; but

* to be devoid of the least understanding of the organisation of the government of Venice, both ancient and modern.

† popular traditions deserving to be believed only by old women and children.

he had prepared a fourth edition of his book, reproducing the full text of Tiepolo's comments and his own responses, and this was published in 1853.

Daru's defence, conducted with condescending patience and deadly forensic skill, was far more impressive than Tiepolo's prosecution, which was alternately pedantic and oratorical. Furthermore, until Tiepolo's text was published in France his work was unknown outside Italy, and even in Italy it had little influence. Ippolito Nievo, for example, relied uncritically on Daru for the historical background of his novel *Confessioni di un italiano*, published in 1857–8.[12] Nevertheless the fourth edition of Daru's book was never reprinted, and it became something of a tombstone to an extinct reputation. During the second half of the nineteenth century the opinion became widespread that in his grim portrayal the Republic had been traduced. A process of exoneration and rehabilitation got under way, stimulated in part by the Rankean revolution. Archival research demolished whole areas of Daru's argument. But antedating the research, and in some measure determining its outcome, was a loss of faith in received wisdom. Incredulity was in the air, affecting the atmosphere of both religious and secular inquiry; and rewriting the history of Venice became a part of that comprehensive reordering of the past, that relentless dismantling of legend, that absorbed so much of nineteenth-century intellectual effort.

Léon Galibert, in a history of Venice published in 1847, acknowledged that the forms and procedures of Venetian government had been flagitious in theory. However, he doubted that they had been so atrocious in practice. He pointed out that the historian of Venice was confronted by a striking anomaly. A system of government that appeared monstrous and terrible had nevertheless been popular and had to its credit an extraordinary record of success and distinction. 'Sans cesse', he observed, 'on s'étonne qu'avec un tel système d'oppression et de tyrannie, il ait pu accomplir au dehors tant d'importantes conquêtes, et qu'au dedans les sciences et les arts aient pu y prendre de si magnifiques développements.'* No doubt many failings were attributable to the Venetian aristocracy—'mais ces reproches ne sont-ils pas inhérents à toutes les aristocraties?'[13] In 1855 Michelet made amends for the libels of previous French historians. In his *Histoire de France* Venice was reinvested with the traditional republican virtues—tolerance, prudent and economic management, concern for public welfare, and freedom from the extortion and favouritism that were inseparable from monarchy. The

* It is a matter of constant surprise that with such a system of oppression and tyranny it was able to effect so many important conquests on the external front and that on the domestic front the sciences and the arts were able to develop so magnificently.

terror had been a shadow—'on a trouvé peu de sang'.[14] As for the notorious Venetian prisons—'qu'est ce, grand Dieu, que les *plombs* et les *puits* dont on parle toujours, en comparaison des Bastilles, des Spielberg, des Cronstadt dont les rois ont couvert l'Europe?'* Baschet, likewise, protested against the vilification that Venice had been subjected to. 'L'histoire, qui aujourd'hui se montre infatigable dans la recherche de la vérité, doit s'imposer la réparation d'erreurs si grandes.'† Charles Yriarte, writing in the early years of the Third Republic, reckoned that France could do worse than adopt Venice as its model. He depicted a state whose wise and professional patriciate, whose impartial rule of law, and whose politics of anonymity had all been exemplary. In Venice despotic caprice and passion had held no sway. He told his countrymen that they should adopt this rule of renunciation ('cette abnégation doit être désormais la loi de notre vie').[15]

When Mas Latrie raised the subject of political assassination, in the *Revue Historique* in 1882, Professor Vladimir Lamansky of St Petersburg entered the debate. He published a thousand-page volume of closely printed documentation which illustrated the use made by Venice of this instrument of statecraft in its dealings with the Greeks, the Slavs, and the Ottoman government. Lamansky's purpose was not to vindicate the Venetians. It was to vindicate the Slavs and to endorse Russia's destiny as the saviour and champion of a civilization superior to that of the decadent West. He reckoned that his revelations would undermine 'la conviction invétérée chez le gros public que l'Européen doit oublier son passé historique et n'étudier que celui de l'Empire de Byzance et de l'Empire de Russie pour comprendre et connaître jusqu'à quel degré de barbarie et de bassesse pouvait parfois descendre l'humanité de notre ère chrétienne'.‡ However, it was an effect of his book to exonerate Venice from the charge of egregious infamy, because it demonstrated that Western Europe generally had been addicted to the practice of political assassination. 'Quant à la moralité', Lamansky concluded, 'le Conseil des Dix et le gouvernement de Venise comme tout autre, a toujours représenté la morale moyenne de l'époque.'§ Furthermore his researches in the Venetian archives convinced

* What in Heaven's name were the *piozzi* and the *piombi* in comparison with the Bastilles and the Spielbergs and the Cronstadts with which the kings covered Europe?

† History, which is nowadays indefatigable in the pursuit of truth, must make amends for such gross errors.

‡ the conviction inveterate among the public at large that, in order to understand and know to what depths of barbarism and baseness humanity in our Christian era can sometimes sink, the European must forget his own historic past and study only that of the Byzantine Empire and the Russian Empire.

§ As for morality, the Council of Ten and the government of Venice represented, like any other, the average morality of the time.

him that the government of the Republic, though heinous in theory, had in practice been but a feeble tyranny. The very frequency and severity of its penal decrees implied an inability to enforce its authority, and muddle rather than deadly efficiency was suggested by its conflicting enactments and overlapping jurisdictions. The Council of Ten, so formidable in legend, seemed from the archival evidence to have impeded rather than assisted the reign of terror. Instead of ruthless and highly trained professionals, Lamansky discovered meddling amateurs who, though more adept than executive agencies elsewhere, nevertheless clogged up government with superfluous paper and red tape, while trying in vain to keep their plots and stratagems secret:[16] 'Les empoisonnements et autres lâchetés du même genre, auxquels le Conseil était enclin à avoir recours contre les ennemis politiques . . . tout en se cachant dans de mystérieuses ténèbres, réussissaient, par bonheur, moins souvent qu'ils n'étaient divulgués au grand jour, pour l'opprobre de la République.'*

In Italy the rewriting of Venetian history became a mode of national revival. It meant liberating territory of the mind from foreign occupation; reasserting the right of Italians to inherit their own past. So a new version of the history of the Republic was made a part of the agenda of the Risorgimento. The Venetian revolutionary government of 1848–9 granted a Jewish scholar from Trieste, Samuele Romanin, unrestricted access to the archives; and the result of his labours, a copiously documented history of the Republic, began publication in 1853. It reached its tenth and final volume in 1861, the year of his death. Romanin's was the work that succeeded Daru's as the magisterial survey of Venetian history. It purported to demonstrate, by reference to unimpeachable archival sources, the purely fanciful character of much of that lore and tradition which Daru had been supposed to validate. The rule of the Council of Ten, it transpired, had been neither arbitrary nor ruthlessly cruel; and the popular belief that it had conducted its business in a dim room hung with black was so much gothic legend. The so-called 'statutes' of the Inquisition of State were proved to be forgeries by the discovery of other regulations, which were indisputably genuine and which bore no resemblance to those quoted by Daru.[17] One by one the accusations were either modified or demolished, and the notion of the evil Republic, the shame of the civilized world, was consigned to the growing scrap-heap of discarded fable. Subsequent research supplemented Romanin's prodigious work of revision, and in the eyes of the Risorgimento

* The poisonings and other cowardly acts to which the Council was inclined to have recourse against political enemies, all the while concealing itself in mysterious shadows, fortunately succeeded less often than they were exposed to the daylight, to the disgrace of the Republic.

generation the Venetian *ancien régime* acquired a new and refulgent prestige. It became, in the words of the Venetian writer Francesco Zanotto, 'the marvellous and meritorious Republic of Saint Mark' ('la meravigliosa e benemerita Repubblica di San Marco')[18]—a paragon of paternal and efficient administration, whose virtues were exceptional but whose vices were not. Zanotto investigated in some detail the physical conditions endured by prisoners in the *pozzi* and *piombi*, and produced documentary evidence to illustrate 'la mitezza e la clemenza' of the Republic's jail regime. Mildness and mercy were less apparent in his own inclinations, however. He left his readers in no doubt that even the *pozzi* and *piombi* of legend would have been too good for libellers like Daru and Fenimore Cooper.[19] In 1871 the *Archivio Veneto*, a journal under learned patronage, was founded in order to make known 'la storia vera di Venezia, in luogo della tradizionale, che si è andata di secolo in secolo ripetendo'.* Among the earliest contributors was Rinaldo Fulin, who was one of the first historians to consult the proceedings of the Inquisition of State. Fulin realized that the case for the defence was at risk from overstatement and counter-myth, so he substituted the measured language of the scholar for the indignant rhetoric of the patriot. He established that the Inquisition's repertoire of punishment had indeed included poisoning and drowning, and that the *pozzi* had been in use until the last days of the Republic. Romanin had erred in claiming that they remained empty after the building of the new prison in the seventeenth century. But Fulin was able to confirm that there was little evidence of severity outside matters of state security.[20]

One of the principal apologists of the old regime was Pompeo Gherardo Molmenti, a Venetian historian who went on to have a distinguished career in Italian public life. His social history of Venice was first published in one volume in 1880, when he was 28, and was based, like Sismondi's study of the Italian republics, on the premiss that there was a direct link between forms of government and national character. But since he found no evidence in Venice of the egregious viciousness deplored by Sismondi, Daru, and others, Molmenti deduced that the tyranny must be illusory too. He argued that the aristocracy's addiction to outrageous pleasure in the last decades of the Republic had been a characteristic common to all European high society in the age of sentiment. The quality of Venetian government on the other hand had been unique. For much of the medieval and modern periods the Republic was the best administered state in Europe. Molmenti accepted the theory of usurpation. The oligarchy had originally been a

* the true history of Venice, in place of the traditional one that has gone on being repeated century after century.

democracy. However, the nobility had atoned for their transgression of popular rights by ruling wisely and benignly. They had allowed themselves no exemption from law, and their justice had been neither arbitrary nor excessively harsh. Most of the horrors associated with their police and penal system were figments of either popular fantasy or foreign malice.[21]

The Anglo-Saxon verdict on Venetian government was modified in accordance with the new understanding that was developing in France and Italy. In 1861 John Stuart Mill numbered the Republic among the few states that were noteworthy for 'systematically wise collective policy and . . . great individual capacities for government'. Its peculiar merits had accrued from the fact that it had been ruled by an oligarchy within an oligarchy—by a bureaucracy in effect. Venice had possessed a corps of professionally trained public functionaries 'remarkable . . . for sustained mental ability and vigour in the conduct of affairs'. In 1877 Acton was still talking of 'a frightful despotism', and charging the Venetians with having followed the notorious 'Maxims' drawn up for their guidance in the seventeenth century by Fra Paolo Sarpi. Yet in the same year Elizabeth Eastlake, writing in the *Edinburgh Review*, suggested that Sarpi's maxims were as spurious as Daru's statutes—part of a sustained attempt by its enemies to discredit the Republic. 'The suspicion', she noted, 'that Venice has been painted blacker than she deserves has been gradually obtaining.' She reviewed the work of the revisionists, and declared herself persuaded. 'It is notorious', she told her readers, 'that Daru was strongly biassed against the Venetians.' She deemed it impossible 'that the most prosperous, liberal . . . patriotic and longest-lived of Christian states should have been the one most iniquitously governed'. In a subsequent article, published in 1889, she trounced Daru as 'impudently false' and commended Molmenti for intelligent candour ('rare in a foreigner'). She rejoiced that fact was now triumphing over fiction and that the Republic was receiving its due for a 'wisdom and . . . coldly reasoned and sternly persistent policy' that had been 'equally removed from the impulse and the sentimentality of modern times'.

Criminal statecraft was largely written out of Venetian history by the later Victorians and the Edwardians. Horatio Brown, it is true, never quite outgrew the strong, simple formula of guilt and retribution. He judged the events of the Fourth Crusade, in which Venice took a leading role, as 'a great crime' requited with 'a lifelong punishment'. This blow at the Byzantine Empire had brought down Nemesis in the shape of the Turks, whose 'establishment at Constantinople, facilitated by the present action of the Republic, left Venice subsequently exposed to a long series of wars, which she heroically sustained . . . but which broke her power, exhausted her strength, and materially contributed to her ultimate ruin'. However, Brown gave up the idea that Venetian government had been essentially malignant. In 1887 he was writing of the aristocracy as usurpers and the Council of Ten as a 'tyranny' ruling with 'a rod of iron'. In 1895 he described the Council as 'a body strictly governed by its own rules [and] constantly changing its component members, who were therefore unable ever to exercise a dangerous abuse of their power'. William Hazlitt, whose bulky history of Venice was published in four editions between 1858 and 1915, implied wickedness and retribution by borrowing imagery from the biblical prophecy concerning Tyre. 'Venice', he intoned, 'now lies in state . . . and the fisherman's net hangs to dry from balconies which have been trodden by all that was illustrious and fair in the wealthiest and most un-European of European centres.' But historical argument soon parted company with rhetorical habit. 'There is scarcely any branch of the inquiry', wrote Hazlitt of Venetian government,

> on which the once-prevailing notions have been so gravely modified, and at the same time so beneficially enlarged, by modern researches. The result has been, on the whole, to acquit the Republic of the gross and absurd calumnies propagated by ill-informed and ill-disposed writers, and its tardy admission to its just place among the great Powers of the Middle Ages and the Renaissance.

There were, he pointed out, 'some respects in which the Venetian laws of the thirteenth century exhibited a greater degree of mildness than the laws of other countries in the eighteenth century'. The French writers who animadverted on the *pozzi* and *piombi* were 'a little forgetful that mankind has never beheld, and hopes no more to behold, anything so barbarous, so degrading, and so loathsome, as the dungeons of the Bastille and the Grand Châtelet as they were even so late as the last quarter of the eighteenth century'. The virtues of Venice were again praised as unique, and its failings condoned as unexceptional.[22]

And it was by its unique virtues, insisted the Harvard historian William Roscoe Thayer, that it should be judged. Thayer, expressing perhaps the disenchantment of New England intellectuals with modern American democracy, rejected the idea that the Venetian aristocracy had usurped popular rights. 'No flourishing democracy was destroyed.' Moreover, the example set by Venice made it impossible to deplore the fact that one had never developed. 'The true historian,' said Thayer, 'though he be a stanch democrat . . . will do full justice to the Venetian oligarchy, even to the point of regretting that no democracy has ever thus far come as near perfection as the political system of Venice came.' He was full of admiration for the

Council of Ten, which he contended had been the epitome of professional government, 'surpassed by no other similar body in sagacity, in ability, and in single-minded devotion to state interests'. Another American writer who vindicated the Venetian aristocracy was the expatriate novelist Francis Marion Crawford. Crawford lived in princely style in Sorrento on the profits of commercial fiction. He launched an avalanche of what Henry James called 'sixpenny humbug'. Towards the end of his life, however, he wearied of pot-boiling and aspired to make his name as a serious historian. His *Gleanings from Venetian History* (1905) is a refutation of the popular view of the Republic and a rebuke to the confraternity of melodrama and romance. Strongly influenced by Romanin, Crawford defended the oligarchy and disparaged 'the innumerable novelists and playwrights' who had filled Venetian history with blood and barbarism. He argued that it was Florence, where democracy had triumphed, that became the victim of tyranny—thus illustrating 'the truth of some of the most important conclusions reached by Plato in *The Republic*'. By now the legacy of Fenimore Cooper was spent. The American novelist who might still have written of the Venetian past with something of the old vision of horror and something of the old regret for a lost cause was William Dean Howells. Howells perceived in Venetian history a tension between a noble, primitive spirit of democracy and a corrupting, enervating spirit of oligarchy. But his study of the Republic, sketched in 1900, was—significantly perhaps—never written.[23]

In Italy the rehabilitation of the totalitarian state proved to be an intellectual rehearsal for its resurrection. The Venetian Republic, now admired as efficient, benevolent, and free from political faction, became a model for those seeking remedies for the disorders and disappointments of the liberal parliamentary state.[24] In Britain the process of reappraisal functioned as a catalyst of neurosis and insecurity, thus adding to the pressures that were discrediting liberalism and making society more intolerant. One of the recurring themes of nineteenth-century writing about Venice was the similarity between modern Britain and the old Republic. It was impossible to study the history of the island-fortress of Venice, with its great naval potency and its wide commercial empire, with its aristocratic oligarchy headed by a titular monarch and with its tradition of resistance against the papacy, and not think of Britain. Venice, to Balzac, was the London of the Middle Ages ('cette Londres du Moyen-âge'). Galibert described how 'pendant plusieurs siècles Venise fut . . . ce qu'est de nos jours l'Angleterre'.* 'There are many

* during several centuries Venice was . . . what England is today.

particulars in which Venetian history may be compared with our own', remarked the Victorian man of letters, Davenport Adams. Disraeli, addressing the House of Commons on world affairs in 1871, expatiated on the 'considerable similarity between the condition of Great Britain and the Republic of Venice'. Thayer made the similarity an integral part of his book. 'There are, indeed,' he wrote, 'so many parallels between Venice and England that I have not hesitated . . . to call attention to them.' And because Venice and Britain were so much alike, it was inferred that the fate of the one must be a premonition of the future of the other. Grim warnings and prophecies abounded. Balzac wrote of the English in Venice as those to whom history was showing their future ('l'histoire montre là leur avenir'). Byron adjured his countrymen:

> in the fall
> Of Venice think of thine, despite thy watery wall!

The painter Turner thought of it often, and in a huge canvas of 1843 (*The Sun of Venice Going to Sea*) expressed a deeply pessimistic reaction to the affinity between the two nations. Ruskin opened *The Stones of Venice* with a pregnant reference to the fate of the maritime thrones of Venice and Tyre, and predicted that England might well be led 'through prouder eminence to less pitied destruction'.[25]

Now it was a consoling feature of the older historiography that, despite obvious and disturbing similarities, it offered an appreciable margin of difference. 'But there are also differences,' Disraeli assured the House of Commons in 1871; and chief among them was the fact that Venice had 'a suspicious and tyrannical oligarchy instead of an open and real aristocracy'.[26] So long as the Venetian state figured as a tyranny, dark with atrocity and iniquity, then Britain was spared the implications of too close a comparability. It was possible to fit the fall of Venice into a simple schema of guilt and retribution from which Britain, with its renowned constitution and exemplary rule of law, was exempt. The popular novelist Margaret Oliphant, when writing her history of the Republic in the mid-1880s, therefore resisted the revisionary trend and continued to insist on the sinister otherness of Venice. She rejected Romanin's exculpation of the Council of Ten. 'His arguments are poor', she declared, 'in comparison with the evident dangers of an institution whose proceedings were wrapped in secrecy and which was accountable neither to public opinion nor to any higher tribunal.' The Venice in her book was 'more rigid than any individual despotism . . . more autocratic and irresponsible than the government of any absolute monarch'.[27] But Mrs Oliphant's tune could not but sound old-fashioned and naïve. Even her compatriots were singing a different song. By now it was so widely accepted that the Republic had combined unexceptional vices with unique virtues, that British writers were retreating from the biblical interpretation of the final catastrophe. They joined the mainstream of historiography, and attributed the fall of the Republic to other causes—causes from which Britain, only too clearly, was not immune.

Prominent among the alternative explanations was the idea that Venice had died a natural death—the death of decrepitude and old age. Yriarte listed 'l'imperfection et la caducité inhérentes à toutes les choses humaines'* among the causes of the fall of the Republic. Molmenti wrote of 'quel corpo decrepito',[28] and maintained that 'Venezia, dopo quattordici secoli doveva pur piegare alla legge fatale di tutte le cose.'† The Anglo-Saxons worked up this notion into an organic theory of birth, growth, and decay. By the time he came to write *St Mark's Rest* (1877–9), Ruskin's mind was moving confusedly in a new direction, though his tone was as dogmatic as ever. He still insisted that 'modifications of . . . policy and constitution' in the history of Venice were produced by 'changes in her temper', by 'changes of [the] spirit'; but he no longer conceived these changes as expressions of collective moral choice. He now explained them in terms of a 'vitally progressive organisation', such as that which caused 'the putting forth the leaves, or setting of the fruit in a plant'.[29] This way of thinking caught on, and the language of historians became almost indistinguishable from that of horticulturalists. 'Had the tree been cut down in its strength,' wrote Elizabeth Eastlake, 'it might have sprung up again; but the sapless shell which looked so fair contained at last only bitter ashes, whence no germs of better life could take root.'[30] In Horatio Brown's first book (*Venetian Studies*, 1887), botany runs rampant. Seeds are sown, germs quicken, fruits ripen, flowers break from bud to blossom, and roots spring into lofty trees. In the preface to *Venice: An Historical Sketch* (1892) the vegetation was cut back to make room for animal analogies. 'Believing that a state is an organic whole,' wrote Brown, 'it appeared to me that the Venetian Republic presented one of the most striking examples of the inception, birth, adolescence, decline, and death of a community which history has to offer for our observation.' So he set out to write 'a biography of Venice'. Before long, however, the jungle was encroaching again, with flowers bursting into full bloom, blossom presaging decay, and death creeping up from the roots before the flower had opened.[31] The Americans adhered more consistently

* the imperfection and the decrepitude inherent in all human things.

† this decrepit body . . . Venice, after fourteen centuries could not but bow to the fatal law of all things.

to the animal analogy, preferring to present Venice in human terms. Thayer wrote that Venice 'lived her own life, grew to full stature, and then slowly passed away, not because she was oligarchic, but because she was mortal'.[32] Crawford insisted not only that 'Venice resembled the living body of a human being', but that it had all the psychological as well as physiological attributes of the female sex:

> She . . . changed from a child to a full-grown woman. Pursuing, or pursued by, the impression of her strong personality as a living creature . . . we may compare her to a woman of divine beauty, yet almost tragically jealous of her own freedom. . . . A woman, in short, possessing a sort of dual nature, aspiring to the dignity of being feared, yet moved by the desire of love.

And 'this beautiful lady, Venice', after a period when she was 'less young indeed, but imbued with a charm more subtle', finally died of 'sheer old age and marasmal decline'.[33] Inevitably, the time would come when Britannia in her turn must succumb to marasma. Macaulay, in his celebrated review of Ranke's *Popes*, had linked the vanished Venetian Republic with a vision of 'some traveller from New Zealand . . . in the midst of a vast solitude, tak[ing] his stand on a broken arch of London Bridge to sketch the ruins of St Paul's.' But since the function of Macaulay's prolepsis was to establish the idea of a vastly remote future (he wanted to suggest the legendary longevity of the papacy), its implication was that Britannia was still young. Much more disturbing were the other theories of decline—the theories that said Venice had fallen victim to the aggressive *Weltpolitik* of European states, and to the toxins and stresses of empire.

'You have a new world,' Disraeli told the House of Commons after the Prussian victories of 1870–1, 'new influences at work, new and unknown objects and dangers with which to cope. . . . The balance of power has been entirely destroyed, and the country which suffers most and feels the effects of this great change most, is England. . . . I cannot resist the conviction that this country is in a state of great peril.'[34] To emphasize his point he reminded the House of the situation of Venice in the early sixteenth century. That was the time when the new territorial monarchies of Europe had confederated against the Republic and worn it out in the wars of the League of Cambrai. All historians of Venice were agreed that this had been the beginning of the end; that the triumphal pageantry of the sixteenth century had masked a decline to the second rank. It needed no great percipience, as a correspondent of *The Times* remarked in a comment on Disraeli's speech, to understand that 'the "agglomeration" of the great states of Russia, Germany, and America' had its precedent in 'the gigantic growth of Austria, France, Spain, and England' in the days of the League of Cambrai, and that Britain's position now was similar to that of the Republic then.[35] Since the collapse of Venice was no longer attributed to crimes, the search was on for errors; and the greatest error of all had obviously been to neglect fleets and colonies in order to play a role in the affairs of the Continent. 'Après que Venise eut acquis des états en terre ferme', wrote Sismondi,[36] 'cette république négligea ses provinces d'outre-mer, son commerce, et sa marine, vraies bases de sa puissance, pour s'engager dans la politique du Continent . . . et elle excita cette jalousie, cette haine profonde et universelle, qui . . . éclata enfin par la ligue de Cambray.'* Daru endorsed Sismondi's diagnosis;[37] and the argument was reproduced in English by Horatio Brown. Both in his *Historical Sketch* and in his chapter in the *Cambridge Modern History* he stressed the disastrous implications of Venice's aspiration to be a Continental power. The lesson for the British was clear. If they wished to avoid the fate of the Venetians, they must remain aloof from *Weltpolitik* and cultivate their naval and colonial strength. And the lesson became compelling as rapid changes in naval technology made old ships obsolete and the loss of naval supremacy more likely. Althea Wiel's history of the Venetian navy, written at the height of the Anglo-German naval rivalry, when the press and the music-halls were clamouring for more Dreadnought battleships ('we want eight and we won't wait'), echoed Newbolt, Kipling, and Baden-Powell, and reverberated with urgent overtones of topicality and warning. 'I have striven', she wrote, 'to set before the general reader the important part the Navy played, for more than a thousand years, in developing the individuality of the Republic, and I have striven to prove how . . . indifference and apathy as to the maintenance of the Navy was the cause of the downfall of that city.' The great days of the Venetians had been their days of 'splendid isolation'. It was then that they had 'fulfill[ed] their destiny as a conquering race, the owners for hundreds of years of the supremacy of the sea'. Venice's 'dominions beyond the sea' had been crucial to its prosperity. They had formed 'the stay and support of the mother-country'; and decline had begun when this empire and the navy that served it were neglected for the sake of territory and influence on the mainland of Italy. The discovery of America and of the Cape route to the East proved disastrous to Venice only because Venice had proved unfaithful to itself.[38] Writing at the same time, Edward Hutton trumpeted the message of

* After Venice had acquired territories on the mainland, the Republic neglected its overseas possessions, its trade, and its navy, the true foundations of its power, in order to involve itself in Continental politics . . . and it excited that jealousy, that deep and universal hatred, that . . . finally erupted in the League of Cambrai.

Venetian history in an even louder incitement to naval supremacy. In his guidebook *Venice and Venetia* (1911) the Republic becomes a surrogate for Britain, and all the complexity of its rise and decline is subsumed into a simple formula, chanted like a slogan again and again: 'Command of the sea'.

Unpopular and insecure, Britain in the late-Victorian period sought and found in empire a solace for anxieties and a compensation for diminished prestige. The Low Imperialism of the mid-Victorian era—the reluctant imperialism of merchants and missionaries convinced that the Empire was a burden (albeit a burden of honour)—gave way to the High Imperialism of the proconsuls and viceroys, with its megalomania, its hysteria, and all that cultish paraphernalia of rite, pageant, and anthem by which the British persuaded themselves and endeavoured to persuade the world that their Empire was unassailable, ordained alike by the laws of God and the laws of Darwin. The Empire was reassurance; it was salvation; but it was also peril. No lesson from history was as unequivocal and as resistant to reinterpretation as that. Its clarity was such that Professor Sir John Seeley argued that the British Empire was not really an empire at all. 'It is much less dangerous to us' he explained, 'than that description might seem to imply. It is not an empire attached to England in the same way as the Roman Empire was attached to Rome; it will not drag us down, or infect us at home with Oriental notions or methods of government.'[39] Seeley undoubtedly had in mind Gibbon's 'corrupt and opulent nobles of Rome', who had 'gratified every vice that could be collected from the mighty conflux of nations'. *The Decline and Fall of the Roman Empire* had planted in the Anglo-Saxon consciousness an enduring awareness of the risk of moral and physical contamination that a hardy race incurred when it encountered soft climates and hypertrophied societies; and the histories of Venice had reinforced the sense of danger. They had depicted another race of conquerors debilitated by the same contagion. 'Elle produisit une révolution dans les mœurs', wrote Daru of the Venetian occupation of Cyprus.

> Celles des Cypriotes étaient extrêmement corrompues; le climat de cette île, toujours mortel aux vertus austères, les jouissances de la mollesse et de la domination, la facilité d'acquérir des richesses, attirèrent les nobles vénitiens et en firent des satrapes voluptueux qui rapportaient ensuite dans leur patrie l'habitude de l'indolence et des plus monstrueux dérèglements.*

* It brought about a revolution in morals. Those of the Cypriots were extremely corrupt. The climate of the island, always fatal to austere virtues; the delights of softness and domination; the ease with which wealth was amassed—all attracted Venetian nobles, and turned them into luxury-loving satraps who subsequently carried back to their homeland the habit of indolence, as well as perversions of the most monstrous type.

By the eighteenth century such influences had produced 'un désordre scandaleux' in public morals.[40] Venice contained 'toute une population élevée dans la plus honteuse licence'.*

Subsequent historians expanded the scabrous theme. Lamansky stressed the corrupting influence of Oriental empire on financial and commercial ethics. He maintained that the decline of Venice had originated in the fortunes rapidly made by aristocratic families in the Levant. These aroused a general fever of greed ('la fièvre du lucre'), and the Venetian dominant class gratified by unscrupulous means a lust for easy money. The Republic experienced a crisis of dishonesty and unfair dealing, which was far more significant than the discovery of the Cape route as a deterrent to foreign merchants. Venice lost its prosperity because it lost its good character.[41] Immorality in private life was investigated by Molmenti, and his revelations prompted Hazlitt to write of 'orgies which would have reduced Juvenal to an insipid trifler and left Rabelais and Aretino little to add'. He noted 'a predominant and ineradicable sensuality recalling the worst Greek and Roman traditions'.[42] The memoirs of Casanova, once they had been authenticated by Fulin, Baschet, Molmenti, and other scholars, substantiated the idea that Venice in its later years had been a sink of debauchery. John Addington Symonds remarked that the memoirs were 'almost precluded from general use by the nature of their predominant preoccupation'. The memoirs of another eighteenth-century Venetian, the playwright Carlo Gozzi, were less scurrilous; nevertheless Symonds still felt compelled to expurgate before publishing a translation.[43] British students found it impossible to accept the pleas in mitigation of these lapses that were being offered by revisionist historians. Thus Elizabeth Eastlake resisted Molmenti's argument that Venice in its decadence was merely typical of its age. 'The corruption of the best is proverbially the worst', she retorted; 'and the corruption of this hard, practical people . . . offers the picture of a vicious effeminacy unmatched as yet in modern history—feasting, revelling, perfuming, masking, and gambling, with other pastimes best left unmentioned.' If such habits no longer featured as crimes inviting retribution, they registered all the more vividly as errors producing debility. 'Most of [the nobles]', continued Lady Eastlake, '. . . were by this time in a state of bankruptcy and physical degeneration.'[44] Her comment shows how closely the British identified themselves with the Venetians, and how prone they were to read in Venetian experience a prophecy of their own. No topic caused more concern in late Victorian Britain than 'physical degeneration'.

* an entire population brought up in the most shameful licence.

Fear of racial deterioration, first expressed in the 1850s at the time of the Indian Mutiny, intensified with the growth of High Imperialism. It was at the heart of the theories of Social Darwinism and Eugenics, as it was at the heart of the accusations and recriminations that filled the press during the Boer War in South Africa. Statistics provided by the army suggested a decline in the physique of volunteers; military disasters in the opening stages of the war indicated 'inefficiency' in national leadership. 'The production of sound minds in healthy, athletic, and beautiful bodies', urged the propagandist Arnold White, 'is a form of patriotism which must be revived if modern England is not to follow ancient Babylon and Tyre.'[45] In 1902 the British government set up an Interdepartmental Committee to investigate physical deterioration. In 1906 Robert Baden-Powell launched the Boy Scout movement to promote healthy outdoor life and sexual abstinence among the young. Both were mindful that in the Darwinian dispensation extinction was the penalty not of wickedness but of weakness; and both were mindful too, no doubt, that the craven capitulation of Venice to Napoleon seemed both to replicate the capitulation of Rome and to anticipate the fate of Britain, so often identified as either the modern Rome or the modern Venice.

A hundred years after the event, then, the collapse of Venice was being explained in terms of degeneration; degeneration was being explained in terms of decadence; and decadence was being excused as a general rather than a specific shortcoming. This interpretation was intellectually precarious, because the concept of 'decadence' was itself now subject to revision. *Kulturgeschichte* had made the categories and judgements of *Staatengeschichte* seem inadequate; Aestheticism had divorced value from ethics and made sensation seem more important than action. So as seen from the *fin de siècle* the Venetian past contained more history than histories of constitution and statecraft could expound, and the Venetian 'decadence' was looking ill-adapted to traditional presentation. The idea of eighteenth-century Venice as a feeble epilogue, requiring to be excused, was the product of a mentality that defined history as past politics and politics as present history. There is evidence of readiness to overturn these preconceptions in the return to critical favour of Giambattista Tiepolo. Hitherto dismissed as a washed-out imitator of Veronese, as artist-in-residence to a declining state and purveyor of ballet-scenes to an effete clientele, his reputation at the turn of the century revived and soared. Maurice Barrès published a eulogistic essay in 1889; John Addington Symonds did likewise in 1893. A major bicentenary exhibition was held in Venice in 1896; and then there appeared full-scale reappraisals by Molmenti, Henry de

Chennevières, and Heinrich Modern.[46] Tiepolo was now reclassified as a great master, startlingly original and modern in his *plein air* rendering of colour, light, and atmosphere. There was still a tendency to separate the man from his age. Symonds argued that Tiepolo had been great in spite of 'the *barocco* taste of the Italian decadence'. The English critic Carew Martin called him 'a stranded survival of the exuberant Renaissance'. But the fact was that Tiepolo had been too famous and popular in his own day, too obviously at one with the life and taste of his time, to figure convincingly as an exception. He had belonged unequivocally to the *settecento*; so when he was revalued, it had to be revalued too.

Rococo Venice acquired new charisma during the last decades of the nineteenth century. The notion was fostered that the Republic had not expired in exhaustion and sterility, but had been cut off in the full vigour of enduring and perverse vitality. Writers and artists visualized a fantastic city where artifice was natural and nature artificial; where the exception was normal and the norm exceptional. The legend took root of the apotheosis of pleasure; of unending carnivals, fêtes, masquerades, and song; of immunity from what Barrès called 'the hypocrisy of prudes, the narrowness of fanatics, and all the banalities of the majority' ('l'hypocrisie des austères, l'étroitesse des fanatiques, et toutes les banalités de la majorité'). It featured in the pictures that Giacomo Favretto painted under the inspiration of Tiepolo, and in the plays that Hugo von Hofmannsthal wrote under the inspiration of Casanova—*Der Arbenteurer und die Sängerin* (1898), and *Cristinas Heimreise* (1909).[47] It featured too in the extraordinarily accomplished study of eighteenth-century Italy that Violet Paget ('Vernon Lee') published at the age of 25 in 1881, after devoting her adolescent years to studying Goldoni, Gozzi, Metastasio, and other forgotten luminaries of Italian drama and music. Vernon Lee dissociated 'decadence' from degeneration, by writing of decadent Venice in terms not of decay but of brilliant survival. It had preserved 'the licence, the practical spirit, the incredulity, the magnificence, fancifulness, and splendid cynical corruption' of the Renaissance. Here hedonism had not been suffocated by the Counter-Reformation. 'The Spaniards and Jesuits had not been at Venice and covered all things with the dirty white monotony of their social and moral whitewash.'[48] The French historian Philippe Monnier, likewise, wrote of 'an enchanted city, a wonderful, mad city of masks and serenades, of amusement and pretence', where the arts of painting, music, and theatre had converged in a supreme moment of civilization, and enjoyment had spanned the whole compass of sensation. In Venice at this time 'the indulgence shown to vice, the absence of scandal and hypocrisy, the extraordinary frankness' had made it seem that 'the capacity for indignation were lost, the power to blush obsolete'. Nowhere else in Europe had morality and conformity been so powerless; nowhere had existence been so vivid and so intense.[49]

In these presentations the Venetian past is fulfilling a new function. It is no longer being used to illuminate the present. It is being used, rather, as a mode of protest against it. For the lessons of the revised historiography had been only too thoroughly learnt. Bismarckian and late-Victorian society conspicuously failed to dissociate degeneration from decadence. In Britain, in fact, there was a concerted effort to eliminate the one by punishing the other. So while the Venetian Inquisition was being portrayed as a pantomime tyranny, mocked and disobeyed, there was brought into existence a British inquisition, whose authority was peremptory and whose writ ran far beyond public life. 'Mais tout cela', wrote Fritz Hohenlohe of the Venice of Tiepolo, 'est si loin, si loin, que les échos de ces fêtes joyeuses ne parviennent presque plus jusqu'à nous.'*[50] And nowhere was the sense of distance so great, or the sound of revelry so faint, as in late nineteenth-century Britain. Here the era now began when whole areas of artistic activity and private behaviour where blighted by censorship, delation, and criminal prosecution. An atmosphere of guilt, shame, and fear was created by legislation such as the Criminal Law Amendment Act of 1885, and by agencies such as the National Vigilance Association and the National Council for Public Morals. All registered a growing determination to counteract degeneration by traumatizing the collective psychology.[51] Pornography, prostitution, and homosexuality were suppressed, and after the ritual destruction of Oscar Wilde any idea of comparability between Britain and Venice again seemed impossibly far-fetched. The laws and taboos that gagged John Addington Symonds and obliterated the memory of half his life were symptoms of an attitude that had no currency in the city of Tiepolo and Casanova. Venice was once more antithetic and remote—though this time it had the remoteness not of a nightmare tyranny but of a magical dream. Not until the 1960s would London rediscover itself in the image of eighteenth-century Venice. It was only after the British Empire was dead, and the values that had supported it buried, that Britain could match the metropolis described by Virginia Woolf as 'the playground of all that was gay, mysterious, and irresponsible'.[52]

* But it is all so far away, so far away, that the echoes of those joyous revelries hardly reach us any more.

# · PART THREE ·

## *Beyond History*

## · VI ·

# *Time's Ruin*

In Venice history had stopped. Behind the activity there was always a stillness; after the music, always a silence. 'There is no city in Europe', wrote Arthur Symons, 'which contains so much silence as Venice.' In Byron's Venice the old music has ended:

> In Venice Tasso's echoes are no more,
> And silent rows the songless gondolier;
> Her palaces are crumbling to the shore,
> And music meets not always now the ear . . .

And when, in Browning's Venice, the ghostly strains resume, there is no one left to whom their language does not sound strange and obsolete:

> What! Those lesser thirds so plaintive, sixths
> diminished, sigh on sigh,
> Told them something? Those suspensions, those
> solutions—'Must we die?'
> Those commiserating sevenths—'Life might last!
> We can but try!'

History had stopped—but time continued; and evidence of time's harm was so magnified and multiplied by an atmosphere empty of great events, that the idea of a dying city became one of the most potent obsessions of the European and American imagination.

The premonition of doom was expressed in two visions of destruction, both suggested by the age-old necessity to protect the city from subsidence, sedimentation, and flood. One was the vision of inundation, which was especially resilient since it contained elements of both Burkean sublimity and biblical retribution. Byron, Shelley, Samuel Rogers, and Tom Moore had all helped to popularize the idea of a watery grave, and the Italian historian Carlo Botta predicted in 1826 that Venice would soon be a heap of ruins half hidden by the sea. Subsequently, it was common for visitors to imagine, like Dickens, the water of the lagoon 'waiting for the time . . .

when people should look down into its depths for any stone of the old city that had claimed to be its mistress'. Few conceits were more overworked in Victorian and Edwardian travel-literature than that of the Adriatic claiming his bride, and few were rendered in deeper shades of purple. 'Gradually he her immortal lover is gathering her into his embrace', wrote Edward Hutton in 1911.

> Soon he will kiss her on the mouth and cleanse her from all the abominations that we have made her suffer. . . . She is thinking of her husband the sea, and of her destined bridal bed. . . . All the spoils of the splendid ships, all the beauty of his prey . . . he will lavish upon her, and every night he will deck her with innumerable stars. Ropes of seaweed, opalescent and rare, will sway like beautiful snakes in her hair, banners woven by the secret sway of the sea shall float from the tall *campanili*.[1]

The vision of destruction by water was as old as Venice itself, and originally it had been a vision of violence. The sea had always been a threat as well as a safeguard to the city, and one of Tintoretto's pictures in the church of Madonna dell'Orto shows palaces and churches drowning after the tempestuous Adriatic has breached the *lidi* and *murazzi*, the natural and artificial defences of the lagoon. This was the annihilation envisaged by Ruskin on the first page of *The Stones of Venice* (1851): 'I would endeavour to trace the lines of this image before it be for ever lost, and to record, as far as I may, the warning which seems to me to be uttered by every one of the fast-gaining waves that beat, like passing bells, against the STONES OF VENICE.' John Addington Symonds, likewise, dreamed of Venice overwhelmed by the sea: 'I saw the billows roll across the smooth lagoon like a gigantic Eager. The Ducal Palace crumbled, and San Marco's domes went down. The Campanile rocked and shivered like a reed. All along the Grand Canal the palaces swayed helpless, tottering to their fall . . .'[2]

However, a different mode of drowning was now also foreseen. Venice would suffocate not in sudden catastrophe, but by inches. In 1810 French engineers excavating the *broglio*, or patricians' promenade, in the lower arcade of the Ducal Palace, discovered that the missing bases of the columns were below the level of the pavement, which had been raised by some 38 centimetres over a period of five centuries to compensate for subsidence of the soil. This discovery made the idea of sinking an essential part of the conception of Venice. In 1852 Dr Thomas Burgess, the celebrated climatotherapist, wrote that Venice was 'doomed . . . to sink under the waves of the Adriatic some sixty years hence, and leave no trace behind'. Expert opinion postponed the evil day; but subsequent excavation confirmed that subsidence was occurring at the rate of about 9 centimetres a century. By the early 1900s half the stone podium on which the huge

Campanile of San Marco was built had disappeared beneath the level of the pavement, and several famous buildings appeared to be under threat. It was reported that the campanile of the Frari had sunk some 18 inches, dragging with it one side of the church, and that subsidence had so far endangered SS Giovanni e Paolo that it was unlikely the building could be saved.[3]

By this time the expectation that Venice as a whole would disappear under the sea had been challenged by a different prognosis. This was the 'high and dry' theory, which prophesied that the sea would abandon Venice, not engulf it. Again, the idea was old. The problem of sedimentation had teased Venetian engineers since the sixteenth century, and Addison had repeated in 1701 the widely held belief that the city would soon be on *terra firma*. The Revd John Eustace, in his often-reprinted guidebook of 1812, predicted that Venice, like Ravenna, would soon be surrounded by sand. 'The [Venetian] Republic', he explained, 'expended considerable sums in cleansing the canals that intersect and surround the city, in removing obstacles, and keeping up the depth of the waters, so necessary for the security of the capital. The interest of a foreign sovereign is to lay it open to attack.' The Austrians in fact were never guilty of such Machiavellian negligence. Nevertheless they did, inadvertently, enhance the threat. In 1840, with the object of protecting the province of Padua from floods, they had diverted the river Brenta into the Venetian lagoon, thereby causing serious sedimentation and swamping in the area of Chioggia. In 1875 *The Times* discussed the problem in a long article and warned that 'the day must come . . . in which the sands of the sea and the deposits of the rivers shall choke up the lagoons, obstruct the canals . . . , swamp Venice, and either kill by wholesale or compel to flight its fever-stricken population'. The vision was of a malarial terrain like the Pontine marshes or the plain of Paestum, and the prediction was that without immediate and drastic action 'the blotting out of Venice from the number of existing cities [was] a calamity sure to occur within a calculable and by no means distant period of time'.[4]

That note of concern was portentous. Hitherto the sentence of death that rested on Venice had not registered as a 'calamity'. Protest and lamentation, later to become so strident, were muted and often absent in early nineteenth-century comment. There were two main reasons for this. First, Venice was still seen as a guilty city, blighted by God's punishment. Commiseration would therefore have been tantamount to condoning iniquity and impugning divine justice. 'Mourn not for Venice,' commanded Moore,

> for vanish'd too,
> (Thanks to that Pow'r, who, soon or late,
> Hurls to the dust the guilty Great,)

Are all the outrage, falsehood, fraud,
The chains, the rapine, and the blood,
That fill'd each spot, at home, abroad,
Where the Republic's standard stood.

His injunction was widely obeyed. 'One cannot wisely regret, indeed, the extinction of its wicked Republic,' agreed the early Victorian ecclesiologist Benjamin Webb; and the Revd Henry Christmas, a seasoned Mediterranean traveller, refused to weep over the remains of the Serenissima. 'To see a city of such marvellous beauty . . . really decaying before our eyes would be most melancholy, if the whole aspect of the place did not forbid the sensation.'[5] The Anglo-Saxon impeachment of Venice was doubly severe, because in the lexicon of Protestant morality architectural extravagance meant evil. Milton had made opulence an attribute of Pandemonium, the palace built by a fallen angel for the glorification of Satan:

Anon out of the earth a fabric huge
Rose like an exhalation . . .
Built like a temple, where pilasters round
Were set, and Doric pillars overlaid
With golden architrave . . .
Not Babylon,
Nor great Alcairo such magnificence
Equaled in all their glories.

That vision of hellish art beside the Stygian lake exerted a subliminal but powerful influence on Anglo-Saxon attitudes towards luxurious cities. It had much to do with British intolerance of places like Delhi and Lucknow; and it was clearly at the back of Samuel Rogers's mind when he versified his impressions of Venice:

among the ocean waves . . .
Rose, like an exhalation from the deep,
A vast metropolis, with glistering spires,
With theatres, basilicas adorned;
A scene of light and glory, a dominion,
That has endured the longest among men.

The Miltonic sonority of those lines from *Italy* conveys a meaning at odds with their rhapsodic inflexion. By borrowing both his scansion and his diction from *Paradise Lost* Rogers in effect identified Venice as a precinct of Satan and of Belial. Another Venetian poem, Arthur Clough's *Dipsychus*, which dates from the early 1850s, does not derive so directly from Milton. Technically it is resolutely anti-Miltonic, and its Satan belongs to *Faust* rather than to *Paradise Lost*. Nevertheless, the Puritan equation is still intact. It is in the opulent sea-city that the Mephistophelian voice is heard, goading the modern youth into existential anguish and then tempting him with consolations of the flesh:

> There was a glance, I saw you spy it—
> So! Shall we follow suit and try it?
> Pooh! What a goose you are! Quick, quick!
> This hesitation makes me sick.
> You simpleton! What's your alarm?
> She'd merely thank you for your arm.

Secondly, there was the conviction that the true beauty of Venice was an attribute not of the mortal city, but of immortal nature. 'Nature doth not die', wrote Byron; and for him and his generation it was to nature that the city owed its most enchanting manifestation. Turner's Venice is a distant vision created by atmosphere and light. Its architecture is insubstantial, and freely modified and transposed. Shelley and Rogers likewise reserved their admiration for a city that was distant, luminous, and elusive. Seen from a closer viewpoint, Venice was not beautiful. Its parts were inferior to the whole. Venetian architecture was impressive in mass and outline, but in detail it was disturbing in a way that suggested something other than beauty. The architect Joseph Woods wrote of 'the impression which everybody feels and nobody can tell why'. 'Singular . . . affecting the mind greatly as the work of some unknown people,' commented Rogers. 'Strange . . .', murmured Frederick Faber. 'Strange . . .', echoed Henry Gally Knight. 'That strange cathedral!' exclaimed Richard Monckton Milnes. 'Édifice étrange . . . étrange édifice,' Gautier repeated, summing up first San Marco, then the Ducal Palace. In their perplexity, visitors tried to decide whether they were contemplating something less than beauty, or something more. Woods decided that, in the cases of the basilica and the Ducal Palace at least, it was something considerably less—in fact, that it was ugliness. The young William Gladstone, in Venice in 1832, reckoned that it was less, but not that much less. 'Venetian buildings', he wrote, 'have in general little more than quaintness or profusion of decoration to recommend them.' The equally young Arthur Stanley, eight years later, likewise spoke of 'quaintness'—but then added to the confusion by qualifying it as 'sublime', which in Burke's formulation implied something superior to beauty. However, the quality most often identified was 'magnificence', which commanded wonder but not approval. It connoted the forbidden gorgeousness that transgressed the limits of style and taste. Venetian magnificence, said Woods, was 'produced by the exhibition of riches and power, and not by just proportion'. Eustace perceived 'gloomy barbaric magnificence' in San Marco. To the historian

Edward Smedley the Ducal Palace was 'the irregularly magnificent pile which still avouch[ed] with proud testimony the ancient majesty of the fallen Republic'.[6] It was assumed that such strangeness and such magnificence must be Oriental, since the Republic was known to have had strong links with the Levant and Asia. Wordsworth had commemorated its imperial dominion in 'the gorgeous East', and it was a commonplace of tourist literature that Venice had, in the words of Disraeli, 'been raised from the spoils of the teeming Orient'. Furthermore the location as well as the history of Venice suggested an Oriental classification. The Orient was Asia; but much more than Asia was Oriental. By the early fifteenth century Byzantium, the polity that had straddled Europe and Asia Minor, both mixing and dividing Latin and Oriental civilization, had been squeezed out of existence by Islamic pressures, and the empire of the Ottoman Turks had absorbed south-eastern Europe and the Balkans. So for 400 years 'the East' had meant east of the Adriatic, and Venice had become, like Constantinople in Byzantine days, the great frontier city. It belonged to two geopolitical hemispheres and it represented racial and cultural heterogeneity. 'The variety of exotic merchandise,' wrote William Beckford from Venice in 1782, 'the perfume of coffee, the shade of awnings, and the sight of Greeks and Asiatics sitting cross-legged under them, made me think myself in the bazaars of Constantinople.'[7]

This affinity and this contiguity provided a ready explanation for anything unusual, and the distinctive medieval buildings of Venice were interpreted in Oriental terms. 'I cannot help thinking St Mark's a mosque', wrote Beckford, 'and the neighbouring palace some vast seraglio, full of arabesque saloons, embroidered sofas, and voluptuous Circassians.' Gautier described San Marco as 'un rêve oriental', and Milnes versified its redolence of the East:

> . . . my friend, whom change
> Of restless will has led to lands that lie
> Deep in the East, does not thy fancy set
> Above these domes an airy minaret?

James Fergusson's verdict on the Ca' d'Oro shows the same propensity. It was, he wrote, 'far more characteristic of the luxurious refinements of the East than of the manlier appreciation of the higher qualities of art . . . on this side of the Alps'. Artists who worked in the city in the earlier part of the nineteenth century, such as Richard Bonington and James Holland, reinforced such assumptions by introducing foreground figures in Oriental costume into their pictures. In fact, in British and in French estimation

Venice was more Oriental than much of the Orient. Samuel Rogers's reference to 'many a pile in more than Eastern pride' was no casual comparison. Venice more fully matched preconceptions about how Eastern cities should look than did the ancient cities of the Levant, whose domestic architecture was almost invariably nondescript or modern. 'L'aspect extérieur', wrote Lamartine of Jerusalem, 'nous avait trompé comme nous l'avions été si souvent déjà dans d'autres villes de la Grèce ou de la Syrie.'* Smyrna he found as prosaic as Marseilles, and even the legendary Damascus was at close quarters unexotic. The street façades of its palaces, built of grey mud and with few windows, reminded him of hospices or prisons. Another French traveller, Gérard de Nerval, complained that in Baghdad and Damascus the visitor saw only buildings of brick and mud, while the painted wooden houses of Constantinople were rebuilt every twenty years. Cairo conformed more closely to the Oriental archetype. 'C'est la seule ville', wrote Chateaubriand in 1806, 'qui m'ait donné l'idée d'une ville orientale.'† But when Nerval went there forty years later, he found that it was being ruthlessly modernized. Disraeli too experienced this failure of recognition. Damascus, he wrote, 'does not possess one single memorial of the past . . . everything has been destroyed and . . . nothing has decayed'. Constantinople was picturesque, but 'it has little architectural splendour, and you reach the environs with a fatal facility'.[8]

The idea that Venice was Oriental probably accounts for a lot of the uncertainty and ambivalence in early nineteenth-century evaluations of the city. Visitors had difficulty in making up their minds about Venice because they had not yet made up their minds about the Orient. During the eighteenth century European hostility to the Islamic Orient had softened. The idea of the Holy Land under infidel dominion had become much less offensive in the sceptical Age of Reason, and the 'Saracen' was no longer the incubus that he had been in the age of the Crusades. Gibbon, rewriting the history of the fall of the Byzantine Empire without Christian partiality, had rehabilitated the Arabs and made the Islamic conquests seem much less like a cosmic catastrophe. Galland had translated the *Arabian Nights* into French, his French had been translated into English, and enlightened Europe had discovered an Orient where art averted evil and life was enriched by innocuous fantasy. By the early nineteenth century the East in its Islamic recension was tolerated and even admired. Goethe said that we are all Muslims; Carlyle promoted Mohammed from the status of impostor

* The exterior aspect disappointed us as we had been disappointed so often already by other towns in Greece and Syria.

† It is the only town which has given me the idea of an Oriental city.

to that of hero; and writers such as Moore, Morier, Scott, and Gautier purveyed a miniature, emasculated Islam that fitted easily into light literature and parlour songs. The East was certainly no longer everything it had been. However, it had not outlived all its traditional associations. Beyond the Levantine periphery was an Orient that remained immeasurably older, vaster, and more fabulous than Classical antiquity. It was an Orient that defied scholarly calls to order and overwhelmed susceptible imaginations—de Quincey's, most notably—with dreams of the monstrous and the chaotic. And even the Islamic East could still seem malignant, alien, barbaric. Here lurked the plague, and turpitude, and sacrilege. Though Islam in the Holy Land was no longer offensive, Islam in Greece was. Goethe said we are all Muslims, but Shelley said we are all Greeks; and much of the opprobrium shed by the Arabs was inherited by the Turks. In the philhellenism of Byron, Shelley, and Hugo, the old hatred of the Saracen lingers and the old horror of things Oriental persists.

So the 'Oriental' attributes of medieval Venice left early nineteenth-century visitors perturbed. In so far as Venetian architecture was redeemed, it was redeemed by its Renaissance and baroque examples. The churches of Palladio and Longhena, and the civic buildings of Sansovino and Scamozzi, all in the recognizable idiom of the European Classical tradition, were easier to assess and to like. However, even these were only faintly praised, since they often flouted the rules of taste and architectural grammar. Longhena's Salute and Sansovino's Libreria were too ornate to command unqualified approval; and one architect, Charles Barry, reproached both Sansovino and Palladio for their solecisms.[9]

Barry's attitude is significant, because he had a notable influence on the architecture of northern Europe during the first half of the nineteenth century. He was a prominent figure in the group that brought the civic architecture of the Italian Renaissance to Paris, Munich, London, and Manchester. It was Barry who introduced the Renaissance *palazzo* into London, with his Travellers' Clubhouse in Pall Mall (1829–31), a building that marked the beginning of the attempt to express architecturally Britain's role as the Venice of the nineteenth century. British admiration for the architecture of Venice itself was muted by the conviction that there was nothing here that modern genius could not match. Venice was not irreplaceable. Its best—that is, Renaissance and baroque—architecture could be replicated and even improved in the new commercial empire of the North. In the *palazzo* façade early-Victorian architects discovered a motif that was much better adapted than the Greek temple front to modern techniques of construction, since it was easily fitted to the grid created by an iron frame.

The astylar façade, which had no pilasters or columns, was especially useful for commercial purposes, since it could compensate for restricted ground space by accommodating four storeys, or even five, instead of the traditional three. Architects such as Barry, Sidney Smirke, and Charles Parnell in London, John Gregan and Edward Walters in Manchester, and William Gingell in Bristol, handling the Renaissance formulae with confidence and even bravura, built clubhouses, warehouses, banks, hotels, corn exchanges, and offices that echoed and paraphrased, but rarely mimicked, the great *palazzi* of Italy—the Pandolfini in Florence, the Farnese in Rome, the Corner della Ca' Grande and the Biblioteca Marciana in Venice. 'One can scarcely walk about Manchester', commented the *Building News* in 1861, 'without coming across frequent examples of the *Grand* in architecture. There has been nothing to equal it since the building of Venice.'[10] Italian Renaissance architecture was even more strongly favoured in France, where it became a visual metaphor not so much for commercial and financial power as for dynastic prestige. Initially Charles Percier and Pierre Fontaine, and subsequently Louis Visconti, Hector Lefuel, and Charles Garnier, led an architectural campaign that turned Paris into a city more Venetian than Venice in its extravagant Renaissance and baroque palaces, sumptuous public squares, and grand thoroughfares.

If the Neoclassicists were critical of Renaissance architecture in Venice, the Neogoths were even more critical of its medieval buildings. The Gothic Revival did not enhance Venice in professional esteem. If anything it lowered it, because the trained eye perceived no true gothic architecture in Venice—or indeed anywhere else in northern Italy. Even Robert Willis, an ecclesiologist with a strong affection for Italian medieval art, admitted this. 'Gothic architecture', he wrote in 1835, 'cannot be said to have flourished in Italy at any time, or to have produced there specimens of beauty and grandeur at all comparable with those which arose under its influence on our side of the Alps.' What one found there was a curious hybrid—'a style in which the horizontal and vertical lines equally predominate'. To the early Victorians 'gothic' designated architecture whose predominant lines were vertical and whose characteristic beauty derived from the exhibition of logical construction and engineering skill. Thus the great Northern cathedrals, which utilized the strength of the pointed arch and the resistance of the buttress, were gothic; but so too was the Crystal Palace. 'No material is used in it', explained James Fergusson, 'which is not the best for its purpose; no constructive expedient employed which was not absolutely essential; and it depends wholly for its effect on the arrangement of its parts and the display of its construction.' Its principle, therefore, was 'the same which . . . animated Gothic architecture'. Judged by these criteria northern Italy in general and Venice in particular were devoid of gothic buildings. In Fergusson's opinion all so-called gothic in North Italy was 'a style copied without understanding and executed without feeling', and he was harshly critical of its technical incompetence. George Edmund Street, who travelled in northern Italy in the early 1850s, likewise concluded that in medieval times 'the Northern architects were developing a much deeper art, and working with much more consummate skill, than were the Venetians'. He found that the Venetians were ignorant of, or indifferent to, the architectonic function of the pointed arch—'employing it often only for ornament and never hesitating to construct it in so faulty a manner that it required to be held together with iron rods'. The deficiency arose because 'they scarcely ever brought themselves to allow the use of the buttress; and this reluctance was a remarkable proof of their Classic sympathies'.[11]

The dissenting voice was Ruskin's. In *The Seven Lamps of Architecture* (1849) and *The Stones of Venice* (1851–3) he argued, with orotund perversity, that Venice contained the quintessence of gothic. As usual, he was deducing a universal theory from a private preference. He loved the Ducal Palace. He called it 'a model of all perfection',[12] by which he meant perfection transcending aesthetic considerations. Concepts like 'beauty', 'sublimity', and 'the picturesque' might explain, but they could not justify, his adoration. He needed to show that the palace was gothic, because only the term 'gothic' carried those connotations of Christianity and moral rightness that his evangelical conscience demanded. So he redefined the style to match the building. The pointed arch, the massy surface, the gabled roof he retained in his inventory of gothic characteristics; but he discarded the buttress, and he added the tie-rod as a legitimate gothic accessory. Buttresses he scorned as a decadent fetish. Northern gothic was destroyed 'by their unnecessary and lavish application'. 'In most modern Gothic', he declared, 'they stand in the place of ideas.' It was no violation of gothic principle to counteract the lateral thrust of gables and arches with thickened walls and corner piers, or even with iron rods where 'the design is of such delicacy and slightness as, in some parts of very fair and finished edifices, it is desirable that it should be'.[13] Out, too, went all evidence of engineering skill. Ruskin defended the Venetian builders against the charge of technical incompetence, but he insisted that gothic architecture was more than an exercise in constructional logic. 'Nothing is a great work of art', he pronounced, 'for the production of which either rules or models can be given.' Gothic architecture was created not by rule and reason, but by the pious and untutored imagination, and it was sanctified by imperfection and variety ('savageness' and 'changefulness').[14]

Since they were designed for the purpose, these criteria could not but qualify the Ducal Palace as gothic *par excellence*. 'The Ducal Palace', declared Ruskin, 'stands comparatively alone, and fully expresses the Gothic power.'[15] The idea won converts, some of whom were influential. Among them was Henry Layard, who was chief commissioner of works when the site and style of the new London law courts were being debated, in 1869. Layard told the House of Commons that the Ducal Palace was a supremely successful example of secular gothic. The professionals, however, remained unpersuaded. Edward Barry, like his father an architect, insisted that the Ducal Palace was not a 'gothic' building in the authentic, Northern sense of the term. 'The vertical principle', he told students of the Royal Academy in 1879, 'is altogether absent, and in its place we have a horizontalism of treatment as marked as in any building of the Renaissance.' In the famous double arcade, which formed the two lower storeys of the building, 'there [was] no thrust of arches against abutments, strengthened by means of piers or buttresses, as in the Gothic manner'. Instead tie-rods were used to counteract the outward thrust of the masonry; and the row of quatrefoil tracery which divided the upper arcade from the great wall of the upper storeys 'ha[d] the appearance of a horizontal band of enrichment'. It was, in fact, 'a beam, or lintel of perforated stone, or rather marble, laid upon numerous supports, to which it communicate[d] the strain of supporting the solid walls erected upon it'.[16] When *The Stones of Venice* was reissued in 1888, the purists were still unconvinced that there was true gothic architecture in Venice. A critic writing in the *Edinburgh Review* claimed that Ruskin was 'merely making out a case for Venetian architecture'. He accused him of 'inventing a theory to defend their weak and unscientific form of building'.[17]

The Neogoths themselves used a modern equivalent of the tie-rod when they adopted iron-frame construction, and they adopted something resembling the medieval Venetian façade for the non-loadbearing walls of the new style of building. As well as its horizontalism, they replicated its flatness, which gave more scope for fenestration and created more floorspace than the deeply recessed and heavily buttressed gothic idiom of the North.[18] So the pointed architecture of a flat, non-buttressed, horizontal type that was developed in the 1850s and 1860s became known as 'Venetian Gothic'. However, its link with Venice was very tenuous. The most prominent purveyor of the style, Sir Gilbert Scott, always insisted that his inspiration came from the medieval architecture of northern Europe. His St Pancras Hotel and St Pancras Station façade were labelled as Venetian Gothic by Charles Eastlake, the Victorian historian of the Gothic Revival; and the *Building News* perceived in them 'a decided Italian feeling, almost too essentially florid and Venetian to commend itself to our critical judgement'.[19] Yet Scott himself asserted that at St Pancras he had divested his style of the Italian element; and a modern authority, Sir John Summerson, while disputing this, admits that the Cloth Hall at Ypres and some Dutch or Flemish town halls are the obvious progenitors of the buildings.[20] Looking at the architecture that was popularly known as 'Venetian Gothic' in Victorian times it is difficult to resist Scott's own judgement. 'Those who say most on this subject', he wrote, 'know little of what is or is not Italian, and if we attempt any deviation from the most familiar types . . . they at once conclude that it is Venetian, though probably it bears but little resemblance to anything in Venice.' The fact is, that the 'High Victorian' gothicists—including Scott, Butterfield, and Street in his early years—had too low an opinion of North Italian work to plagiarize it, and they were too intent on perfecting a modern gothic style to be content with what Scott called 'dull copyism'.[21] They conceded that useful lessons about colour and brickwork might be learnt south of the Alps; but if there is a Venetian prototype behind any of their buildings, it has been solidified, magnified, and bedizened almost beyond recognition. It has been translated from marble to stone; deprived of 'applied colour' (reckoned unsuitable for the British climate) but lavishly embellished with 'constructional polychromy' (nicknamed 'streaky bacon'); capped with steep roofs and precipitous pinnacles; stripped of delicate *passementerie*; and freighted with what the *Building News* called 'a plethora of florid ornament, carving, and colour'.[22]

For most of the nineteenth century, then, the view in British and French cultivated circles was that Venice was a guilty city whose architecture, though transfigured by nature, was in itself nothing especially admirable, and certainly nothing that could not be equalled by practitioners of Renaissance and gothic idioms in the North. Threats to Venice, both natural and man-inflicted, were consequently not matters of widespread concern.

After the surrender of Venice to the French in 1797 Bonaparte ordered the demolition of some seventy churches and convents, and of several dozen palaces, as part of a plan of extensive clearance and redevelopment. The Austrian years, too, were a period of systematic destruction and reconstruction, with peaks of activity in the 1820s and the 1860s. About a dozen churches and as many palaces perished in the 1820s; and in the early 1860s a swathe was cut through the buildings on the upper reaches of the Grand Canal to make room for the railway station. Several famous monuments were razed or mutilated in these operations. Most notably the monastery, *scuola*, and church of the Carità, declared by Goethe to be one of the greatest architectural ensembles of the world, were stripped and gutted prior to conversion into the Accademia; and the convent and *palazzi* of the quarter of Santa Lucia, often painted by Canaletto, disappeared. Some of the narrower canals, or *rii*, were filled in; and two iron platform bridges, built by the English firm of Neville, were thrown across the Grand Canal—one at the church of the Scalzi, and the other at the Accademia. After the incorporation of Venice into the Kingdom of Italy the main concern of the Consiglio Communale was to modernize the city by opening up its dense medieval texture. The first of several new avenues, the Strada Nuova, linking the Campo Santi Apostoli with the Campo Santa Fosca, was begun in 1868; and in 1871 an area was cleared at San Paternian for a new *campo* with a monument to the patriot Daniele Manin. One of the proposals of Count Luigi Torelli, prefect of Venice in the late 1860s and early 1870s, was designed to make the Giardini Pubblici accessible by carriage from the Piazza. He wanted to construct an iron suspension bridge, 900 metres long and 8 metres wide, parallel to the Riva.[23] These and similar measures and suggestions aroused little comment and less objection in the British and French press. Neville's Accademia bridge, one of the most anathematized features of Venice at the time of its demolition in 1933, was praised by the *Illustrated London News* as a 'handsome structure' with 'elegance of form'. It was inconceivable to the early Victorians that their civil engineering could be aesthetically offensive. Fergusson cited the four bridges over the Thames (Westminster, Blackfriars, Waterloo, and London Bridge) as examples of the 'brilliant and rapid' progress achieved in this area of building art.[24]

The looting of Venice by Napoleon, and the transportation of many of its art treasures to Paris, had provoked British indignation. It was at the insistence of Castlereagh that some of these were restored when the city was transferred to Austria in 1815. However, in subsequent years the British themselves connived at pillage. The antiquity-merchants and picture-dealers who proliferated in Venice in the Austrian period did not lack British custom. A lot of what they sold were fakes and assorted junk concocted for the tourist market; but they were also the channel through which much of the Venetian artistic patrimony seeped away. 'Palaces are now daily broken up like old ships', wrote Disraeli in 1837,

> and their colossal spoils consigned to Hanway Yard and Bond Street, whence, reburnished and vamped up, their Titanic proportions in time figure in the boudoirs of Mayfair and the miniature saloons of St James's. Many a fine lady now sits in a Doge's chair, and many a dandy listens to his doom from a couch that has already witnessed the less inexorable decrees of the Council of Ten.

The dealer Antonio Sanquirico, whose emporium of curiosities filled the huge Scuola Grande di San Teodoro in Campo San Salvatore, disposed of the celebrated Grimani marbles and of a canvas by Mantegna (*The Introduction of the Cult of Cybele at Rome*). This was bought by the English connoisseur George Vivian, and is now in the National Gallery. William Bankes discovered that during the siege of 1848–9 'since nobody had a farthing anything might be had for money'. Among his trophies was a painted ceiling from the Palazzo Grimani, attributed to Veronese and Giorgione, which he had installed in the state bedroom at Kingston Lacey. He also picked up two ornamental well-heads, and used them as tubs for bay trees in Kingston Lacey park. Venetian well-heads were highly prized as garden ornaments by wealthy British collectors. George Cavendish Bentinck, scholar grandson of the duke of Portland, had a score of them spread around the grounds of his home on Banksea Island, in Poole Harbour. In 1814 there had been 5,000 well-heads in Venice. By 1856 the number had been reduced to 2,000. Fifty years later only seventeen of the earliest, Byzantine type remained, and half of those were reckoned to be in the hands of antiquity dealers. Rawdon Brown acted as a tout for picture dealers, and negotiated the purchase of a codex of drawings by Jacopo Bellini on behalf of the British Museum. In 1857 the British consul in Venice, acting for the British government, arranged the purchase from the last of the San Polo branch of the Pisani family of Veronese's *La Tenda di Dario*, and diplomatic pressure was used to obtain the permission necessary for its removal to London. Alexander Malcolm, too, took his profit from the trade in antiquities. He exported to England fragments of rare marble removed from the southern façade of San Marco during restoration, and as late as 1886 Enid Layard recorded in her journal that she and her husband had been taken by Mr Malcolm 'to Madonna dell'Orto to see a palace where doors and ceilings were for sale'. For a long time even people of taste and education regarded Venice as a scrap-heap and a quarry. Any misgivings about the ethics of plunder seem to have been dispelled by the knowledge that Venice itself had been adorned with loot, and by the guilt that accrued from its history and Oriental associations. In Oriental cities, as Delhi and Lucknow discovered after the suppression of the Indian rebellion, the British considered themselves exempt from the constraints of civilized behaviour.[25]

Perhaps the most serious threat to Venice in these years came neither from nature nor from the vandals, but from the goths—from the architectural Neogoths, that is, who were just as active in restoring old buildings as they were in erecting new ones. This obsession with restoration was an important variant of nineteenth-century medievalism, and it acts as a reminder—if any reminder were still necessary—that the age of Romanticism was not monopolized by romantic sensibility. In some ways in fact the Gothic Revival signified resistance to the Romanticism that engrossed early nineteenth-century painting and literature. It was much closer to the purism of Ranke's historiography. Neogothic taste for the rigours of 'style' and constructional logic conflicted with the Romantic validation of spontaneity and expression, just as the Rankean insistence on accurate documentation and scientific truth conflicted with imagination and rhetoric. Furthermore, the promotion of gothic as a secular architecture challenged Romantic assumptions about the religious character of the pointed idiom. But the goths were nowhere more anti-Romantic than in their preference for original perfection rather than for ruins, and in their confidence in their ability to recreate that perfection through antiquarian research and the application of the rules of style. This concern to resurrect the architectural corpses of the Middle Ages explains why the concept of revival was so much stronger than that of survival in the nineteenth-century gothic mentality.

In architecture, then, Romanticism favoured ruin, not revival; but the taste for ruins goes back a long way and illustrates another truth that has become a cliché. The age of the Enlightenment was not monopolized by classical sensibility. Ruins had claimed a place in European literature since the Renaissance, when ancient Rome and its relics stimulated curiosity and interest; but it was about the middle of the eighteenth century that they began to exercise an emotive appeal and to command an art and literature of their own. These were a result partly of the existential anguish of an intellectual élite who had emptied eternity of all attributes save time, and history of all purpose save human achievement; and partly of a new aesthetic appreciation of ruins in and for themselves. Diderot found ruins more appealing than buildings in pristine perfection, and this preference was inherited by Mme de Stael, Benjamin Constant, François-René de Chateaubriand, and Victor Hugo. Constant attested to the expressive power of dereliction: 'Les édifices modernes se taisent; mais les ruines parlent.'* Hugo held that a building was only truly complete when time had worked its alchemy upon it.[26] In Venice in 1833 Chateaubriand rediscovered the special beauty of decay and decided that the modern touch was a blight:

Quand on avise la truelle de mortier et la poignée de plâtre qu'une réparation urgente a forcé d'appliquer contre un chapiteau de marbre, on est choqué. Mieux valent les planches vermoulues barrant les fenêtres grecques ou mauresques, les

* Modern buildings are silent; but ruins speak.

guenilles mises sécher sur d'élégants balcons, que l'empreinte de la chétive main de nôtre siècle.*[27]

Restoration, as defined and practised by Gilbert Scott in England and Eugène Viollet-le-Duc in France, was directly antithetical to these feelings and preferences. It meant something much more than straightforward stabilization and repair. It meant completing a concept. 'Restoration' demanded an expert understanding of structure and a perfect command of architectural idioms, and it could lead to a building's looking as it had never looked before. The principle of notional integrity allowed the preservation of high-quality additions and alterations; but in practice anachronisms were ruthlessly minimized.[28]

The assumption that it was within the power of modern scholarship and technical skill both to elucidate a medieval conception and to realize it, obviously did nothing to encourage respect for old workmanship. Consequently, restoration was often extensive and brutal. At Carcassonne Viollet-le-Duc recreated an entire medieval town. In Britain Scott systematically replaced anachronistic but authentic Perpendicular work with his own 'Middle-Pointed' (i.e. Decorated) substitutions. All over Italy, likewise, architects were turning medieval fragments into Neogothic wholes, and giving medieval finishes to buildings that had never had them. In Florence, the uncompleted Santa Croce and cathedral (Santa Maria del Fiore) were given façades; in Milan San Simpliciano and Santa Maria del Carmine were finished; and in Naples a Neogothic façade was substituted for the eighteenth-century front of the duomo. In Venice restorers were let loose first by the Austrian and then by the Italian authorities, both of whom reckoned that a run-down, neglected appearance was bad for trade and tourism. Between 1816 and 1840 over 5 million lire were spent on urban improvements, including the restoration of derelict churches and *palazzi*.[29]

Extensive work was carried out on the basilica of San Marco. The northern façade was under repair for twenty years, from 1843 till 1865, and the southern façade, overlooking the Piazzetta, for ten years, from 1865 till 1875. Much of the original Oriental marble revetment was replaced with newly quarried *verde di susa* and Verona marble; the protruding back altar of the Zeno Chapel, interpolated in the sixteenth century, was removed; and the marble columns and porticoes were cleaned with abrasives. Inside the church, the deteriorating twelfth-century mosaics of the Cappella Zeno

* When one notices the trowelful of mortar and the handful of plaster that an urgent repair has slapped against a marble capital, one is shocked. The worm-eaten planks boarding up Greek and Moorish windows, and the rags hanging out to dry from elegant balconies, are worth much more than the imprint of the miserable band of our century.

were replaced with modern copies, and the cracked and uneven marble pavement of the north aisle was dug up and relaid with mosaics supplied by the Anglo-Italian firm of Salviati. After 1853 the operations at San Marco were supervised by Giovanni Battista Meduna, an architect who specialized in resurrection and refurbishment. He had made a considerable reputation by rebuilding the Fenice theatre after the fire of 1836. He was also responsible for alterations to the Ca' d'Oro, which had been found inconvenient by its new owner, the ballet-dancer Fanny Taglioni. Among other measures Meduna removed the balconies, filled in some windows and opened up others, and demolished the interior gothic staircase. But in Venice restoration at its most drastic was reserved for the Fondaco dei Turchi, the thirteenth-century *palazzo* on the Grand Canal that had been assigned to the Turkish merchant community in the seventeenth century. The treatment accorded to this ruin under the direction of Federico Berchet provides a remarkable confirmation of the perils inherent in the nineteenth-century aspiration to read the minds of medieval builders. Berchet demolished the remains of the Fondaco and then rebuilt the structure not as it had been (proto-Renaissance) but as he thought it had been (Byzantine). The result, still to be seen on the Grand Canal, is a classic instance of falsification in the cause of notional integrity.[30]

On the island of Murano in the Venetian lagoon the twelfth-century basilica of SS Maria e Donato was restored by the young architect and professor at the Brera Academy, Camillo Boito. Brother of the composer and librettist Arigo Boito, he did for Neogothicism in Italy what Scott, Street, and Viollet-le-Duc had done for it in England and France. As teacher and practising architect he popularized the *Stile Boito*, a polychromatic and ornamented idiom that had strong affinities with High Victorian gothic; and in the first half of his career he perpetrated restorations that showed all the rashness and brashness of those in the North. His work at Murano, begun when he was 22, was carried out under the influence of Street, whose book on the architecture of North Italy he had recently reviewed for the Italian press. He demolished the seventeenth-century side chapels, and substituted imitation twelfth-century fenestration in the place of Palladian windows. The ruined façade he rebuilt with brick, terracotta, and ceramic, and embellished with horizontal bands of colour deduced from the geometric decoration of the apse. The result, as in the case of the Fondaco dei Turchi, was a nineteenth-century hypothesis set up in the place of an authentic ruin.[31]

But the triumphalism of these Risorgimento years obscured any sense of loss, and the restorations of San Marco, of the Fondaco, and at Murano were widely welcomed as proof that modern Italy was able and willing to resurrect the art of the medieval past. Discussing the proposal to remove the rear of the Zeno Chapel from the southern façade of San Marco, the *Gazzetta d'Italia* commented in 1868: 'Questa bella modificazione renderà la Basilica un perfetto monumento bisantino senza alcun deturpo, cosa rarissima negli monumenti, i quali soggiacquero per la maggior parte ad aggiunte ed a ristauri che si resentono dell' epoca in cui vennero fatti.'* In February 1876 the eminent art critic Vincenzo Mikelli wrote in the *Gazzetta di Venezia* with firm approval of Meduna's work on the southern façade and suggested further modifications to disencumber the edifice of anachronisms. Educated opinion endorsed this verdict and there was widespread satisfaction at seeing the old buildings cleaned and straightened. The foreign verdict too was positive. Gautier warmly appreciated the work that Mme Taglioni had had done on the Ca' d'Oro. In 1872, following a short visit to Venice, Viollet-le-Duc gave his blessing to the operations at San Marco, and praised Boito for reviving the 'original form' of the Murano basilica. Generally he found that historical monuments were much better looked after in Italy than in France. Judging by responses to Venice, the Romantic love of ruins was very slow to penetrate the Anglo-Saxon mentality. Until late in the nineteenth century there is little evidence to suggest that British and American observers regretted the loss of picturesque decay. To Byron, Venice had been beautiful in spite of, not because of, its ruined state. Mary Shelley reckoned in 1842 that 'the dilapidated appearance of the palaces, their weather-worn and neglected appearance' diminished the attraction of the city; and five years later the American George Stillman Hillard complained that 'an . . . air of careless neglect and unresisted dilapidation [was] everywhere too plainly visible'. The young Disraeli preferred Venice by moonlight because 'the mystic light concealed the ravages of time'. Mark Twain likewise found Venice by day depressing. 'In the glare of day', he wrote, 'there is little poetry about Venice. . . . In the treacherous sunlight we see Venice decayed, forlorn, poverty-stricken, and commerceless—forgotten and utterly insignificant.' It was only 'under the charitable moon' that the sentimental soul could bear to see it. So when in December 1877 the *Building News* applauded the Italians' determination 'to restore to their primitive grandeur the monuments of national history', it was saying what most people in Britain and America had until then been thinking.[32]

* This fine alteration will make the basilica a perfect Byzantine monument free of disfigurement—a very rare thing among ancient buildings, almost all of which have been subjected to extensions and restorations that carry the imprint of the age in which they were made.

Ruskin alone was mortified. It was during the ten years that followed his first visit to Venice, in 1835, that the changes in the city were most rapid and radical, with the building of the railway viaduct, the installation of gas lamps in the streets and iron balustrades on the bridges, and the demolition and repair of numerous buildings. His second sight of the place, in 1845, was therefore a shock and he wrote to his father in 'a state of torment'. He told him too of a nightmare in which his gondola turned into a steamboat—the premonition of a metamorphosis that was to torture him in later years. 'The rate at which Venice is going', he wrote, 'is about that of a lump of sugar in hot tea.' He was now able to compare the ruin wrought by time with the devastation created by man. At Lucca he had seen monuments that were 'beautiful wrecks', touching in their mortality. In Florence and Venice he saw systematic destruction perpetrated in the name of repair and restoration. It is in *The Seven Lamps of Architecture* that the idea of Venice as something uniquely precious—a miracle that could not be reworked, a dream that could not be redreamt—first appears; because it is here that Ruskin lays it down that a gothic building, once ruined, can never be reconstructed:

> It is *impossible*, as impossible as to raise the dead, to restore anything that has ever been great and beautiful in architecture . . . That spirit which is given only by the hand and eye of the workman can never be recalled . . . Do not let us talk then of restoration. The thing is a lie from beginning to end.

The corollary of this principle, that no second Venice could ever be built, was implicit in *The Stones of Venice*, where the message was delivered that architecture was dead. 'The grotesques of the seventeenth and eighteenth centuries . . . close the career of architecture in Europe. . . . From that time to this no resuscitation of energy has taken place, nor does it for the present appear possible.' This was especially true of England, because 'the modern English mind . . . intensely desire[d], in all things, the utmost completion or perfection compatible with their nature'. The modern English mind was a Renaissance mind. It could therefore duplicate Renaissance Venice. The Renaissance palaces of Venice were 'not more picturesque in themselves than the clubhouses of Pall Mall'. They possessed 'no more interest than those of London or Paris'. They had been concocted from 'recipes for sublimity and beauty', and there was no reason why, 'in this age of perfect machinery', they should not be replicated *ad nauseam*. 'It is certainly possible, with a little ingenuity, so to regulate a stone-cutting machine as that it shall furnish pillars and friezes to the size ordered. . . . An epitome, also, of Vitruvius may be made so simple, as to enable any bricklayer to set them up at their proper distances, and we may dispense with our architects altogether.' However, while easy to copy, this architecture was not worth copying. It was 'pestilent art', detritus left by 'the foul torrent of the Renaissance'. Like all his generation, Ruskin perceived and wrote of Venice under the influence of Milton. Its fall had been just retribution for infidelity and pride. He merely shifted the focus of guilt, making the abode of Belial the Renaissance city rather than the medieval one.[33]

Yet Ruskin spoke with two voices. Ruskin the Jeremiah, proclaiming that architecture was dead and that Venice could never be rebuilt, was contradicted by Ruskin the messiah, who preached the resurrection of architectural art and promised a new Venice as the reward of obedience. 'I plead for the introduction of the Gothic form into our domestic architecture,' he announced. 'I believe it to be possible for us not only to equal, but far to surpass, in some respects, any Gothic yet seen in Northern countries.' Adopt the gothic style of thirteenth-century England and France, he commanded, and finish with Italianate detail, 'and the London of the nineteenth century may yet become as Venice without her despotism, and as Florence without her dispeace.' It was only when he saw what Victorian architects understood by Italianate detail, and after he had deeply pondered the conditions under which modern workmen laboured, that the Jeremiah voice prevailed and Ruskin acknowledged the fatuousness of his messianic *alter ego*. By the early 1870s he was blaming himself for the architectural fad which had 'mottled our manufacturing chimneys with black and red brick, dignified our banks and drapers' shops with Venetian tracery, and pinched our parish churches into dark and slippery arrangements for the advertisement of cheap coloured glass and pantiles'. His promise of a new Venice had ended in travesty. 'There is scarcely a public house near the Crystal Palace but sells its gin and bitters under pseudo-Venetian capitals . . .'[34]

Because of this ambiguity, Ruskin has been credited both with creating and with killing the last or 'High Victorian' phase of the Gothic Revival. The first claim was made by Charles Eastlake, in his *History of the Gothic Revival* (1872); the second by Kenneth Clark, in *The Gothic Revival* (1927). Both claims probably exaggerate his influence. High Victorian gothic was already in existence when he published *The Stones of Venice*,[35] and it was already dying when he turned his eloquence against it. His was, in fact, only one voice in a growing chorus of disapproval and disappointment. By the early 1870s Fergusson was castigating the Neogoths. He accused them of producing not architecture but archaeology: academic exercises in a dead language. His strictures were endorsed by the architectural writer and social critic John T. Emmett, who declared that architecture had been created by free craftsmen and killed by professional architects. The architect was a modern phenomenon ('the ancient builders and medieval masters had no knowledge of him') whose main achievement had been to turn the free, expressive workman into a slavish copier of dead men's ideas. Robert Kerr, professor of the arts of construction at King's College London, reckoned that as a result of Neogothic dogmatism architecture had become 'the most unpopular profession of modern times'. Architects themselves experienced a crisis of self-confidence. 'If we copy', wrote William Burges in 1868, 'the thing never looks right [and] the same occurs with regard to those buildings which do not profess to be copies; both they and the copies want spirit. They are dead bodies . . . We are at our wits' end and don't know what to do.' In 1873 Thomas Graham Jackson published *Modern Gothic Architecture*, whose purpose was 'to inquire why it is that the Gothic Revival has not produced those favourable effects on modern art that might have been looked for'. Later he confessed: 'I had begun to doubt whether . . . we modern Goths were not after all pseudo-Goths . . . masqueraders.' After the mid-1870s the attempt to revive gothic architecture for secular purposes was given up. In Britain architects like Norman Shaw and John Stevenson led a retreat to Queen Anne classicism, while Philip Webb tried to escape altogether from the tyranny of style and recreate architecture from first principles and basic vernacular motifs. On the Continent gothicism, and revivalism generally, had by the end of the century been discarded in favour of an ecclectic, ahistorical idiom—*Art Nouveau* in France; the *Stile Floreale* in Italy. The Gothic Revival passed into history and was numbered by the post-Victorians among the appalling aberrations of Victorian taste. 'It produced', wrote Kenneth Clark in 1927, 'so little on which our eyes can rest without pain.'[36]

And as European architects lost confidence in their ability to create a new gothic architecture, so too they lost confidence in their ability to restore the old. During the 1870s restoration in its Neogothic formulation fell out of favour. It was disowned by the older generation and anathematized by the younger. Philip Webb called it 'ruthless refinements of cruelty by make-believers'. John Stevenson, a pupil of Scott, denounced the restorations of Scott and Street as vandalism. In 1877 Webb, Stevenson, and William Morris founded the Society for the Protection of Ancient Buildings (SPAB), whose manifesto denied that it was possible to 'restore' or 'revive' the architectural heritage. As a popular art, architecture was dead. It survived only as an academic exercise, whose achievement was erudite but lifeless forgery. At its touch picturesqueness withered, history vanished, and the unique became commonplace. 'Modern art', the manifesto declared, 'cannot meddle without destroying.' Whether the SPAB was very influential is debatable. It seems to have exasperated as many influential people as it converted. But the ideas it propagated, either because of or in spite of its efforts, became orthodoxy. The late Victorians conceded that their role in the history of architecture was not creation but conservation. As they saw it architecture was, in the words of Frederic Harrison, 'the inheritance which the past [was] bequeathing to the future'.[37]

Ruskin, in his Jeremiah mood, had said all this. He had said it twenty years before, so it is unlikely that he was directly responsible for the collapse of confidence; but the eddies and reversals that were upsetting old evaluations certainly left the educated public more disposed to listen to his message. His reputation benefited from the new reverence for the past, and the campaign for conservation was largely fought in his name. Attempts have been made to identify Ruskinism as a symptom of an 'English disease'; as evidence of a 'decline of the industrial spirit' that was a characteristic of the British ruling class. Yet these insular references are insufficient, because Ruskin's audience was growing abroad as well as at home. In spite of a ban on integral translations, which was not removed until after his death, he was acquiring a European celebrity. The main intellectual currents of the Victorian age flowed from the Continent to Britain; but Ruskin, like Darwin and Wilde, sent ideas in the opposite direction. By the end of the nineteenth century Spanish and German artists and thinkers were acknowledging his importance, and France and Italy, of whose architecture he so often wrote, were counted among his spheres of influence.

Ruskin had been known in France since the 1860s, chiefly as an art critic connected with the Pre-Raphaelites; but in the 1890s his social and architectural theories were being discussed in intellectual circles. A series of five articles devoted to his ideas was contributed to the *Revue des Deux Mondes* by Robert de la Sizeranne in the mid-1890s, and then published separately as a book (*Ruskin et la religion de la beauté*) in 1897. Meanwhile other influential Parisian journals were printing translated extracts from his works; and the first complete translations, of *The Seven Lamps* and *The Crown of Wild Olive*, appeared in 1900, within a few months of his death. He was then discovered by Marcel Proust and entered the mainstream of European literature as the patriarchal exponent of Venice in *À la recherche du temps perdu*. Proust wrote that having read Ruskin 'l'univers reprit tout d'un coup à mes yeux un prix infini',* and he paid homage with two translations—*La Bible d'Amiens* in 1904 and *Sésame et les lys* in 1906. The

* The universe suddenly became again infinitely precious to my sight.

following year he said that he had now given up translating Ruskin partly because everybody was doing it; and by 1912 he was talking about a process of canonization.[38] 'Ceux qui m'ont le plus reproché ma faiblesse à son égard, en font maintenant un dieu sans défauts, sans mélange du périssable.'* In Italy Ruskin became equally renowned. As solicitude for ancient buildings increased and willingness grew to disparage the present as unworthy of the past, he was invoked more and more as an oracle and mentor. In the early 1870s a small group of young Venetian artists and writers adopted his teaching and started the discussions that were to give his name and his ideas wide currency in Italian intellectual circles. As a result of their canvassing, he was elected an honorary member of the Accademia in 1873.

One of these disciples was Count Alvise Zorzi, a painter and scholar who by the time of his death in 1922 had become a leading figure in museum administration in Italy. Zorzi never learnt English, but he became a friend and correspondent of Ruskin after meeting him in the mid-1870s. In 1877, at the age of 31, he made himself famous in Venice first by campaigning—successfully—to save the baroque church of San Moise from demolition, and then by publishing (with Ruskin's financial help) a highly critical account of Meduna's work at San Marco (*Osservazioni intorno ai restauri interni ed esterni della Basilica di San Marco*). This pamphlet was in effect a conservationist manifesto. It denounced current methods of restoration as vandalism, deplored the ruthless modernization of Venice, and pleaded for the removal of the profit-motive from the upkeep of ancient buildings.[39] Another, equally notable pupil was Giacomo Boni, the most distinguished Italian archaeologist and antiquarian of his generation. From 1888, when he was appointed inspector of monuments in the Direzione Generale delle Antichità e Belle Arti, Boni was the government's chief agent for the care of ancient buildings, and he was chiefly responsible for leading official opinion away from the principles of Viollet-le-Duc and bringing it into line with those of Morris, Webb, and the SPAB. He had learnt English in order to read the works of Ruskin, whose acquaintance he had made in 1876, as a 17-year-old student; and he was one of the fifty signatories of a pamphlet of 1882 entitled *L'Avvenire dei monumenti a Venezia*. This admonished Viollet-le-Duc and his followers, but quoted with fulsome approval passages about the mendacity of restoration and the pestilence of the Renaissance from *The Seven Lamps of Architecture* and *The Stones of Venice*.[40] Even Camillo Boito, the Italian arch-Neogoth, became very uncertain about the rightness of much that was being done in Venice in the name of restoration,

* Those who most reproached me for my weakness for him now regard him as a god without fault, without admixture of the perishable.

and began to look for a formula that would reconcile the conflicting claims of archaeology, stability, and aesthetics. Touchy and combative in his patriotism, Boito resented criticism that was flavoured with foreign ideas, and he dissected Ruskin's theories with irony. But his readiness to turn British arguments against the British themselves (he complained that Oxford had been ruined by restoration) shows a fundamental unease with the whole rationale of Neogothic practice, and soon this hardened into rejection. In 1883 he drew up a code of practice for restorers which was adopted by the fourth Congress of Italian Architects in Rome. It was full of compromise and pragmatism, but it endorsed the Ruskinian prescription of clearly dated and undisguised repair in lieu of architectural reproduction. The following year, on the occasion of the Turin Exhibition, he published an indictment of Viollet-le-Duc that might have been written by Ruskin himself.[41] Viollet-le-Duc, he said, provided no safeguard against arbitrariness, 'e l'arbitrio è una bugia, una falsificazione dell'antico, una trappolata tesa ai posteri. Quanto meglio il restauro è condotto, tanto più la menzogna riesce insidiosa e l'inganno trionfante.'*

So the collapse of the Gothic Revival and the rejection of its theory of restoration were vagaries not of British malaise but of European fashion. The transferral of approval from modern to medieval gothic was a characteristic caprice of a European intelligentsia that was forever desiring, and never finding, something new to believe in. As often as the nineteenth-century voice said 'has been', the nineteenth-century ear heard 'not yet'. The agnostic remained pious, and the Romantic rediscovered a beauty that hated movement and cherished line.[42] The aesthetics of expression, and the cults of the picturesque and of what Ruskin called 'shattered majesty', had never eradicated nostalgia for the aesthetics of form and for the ordered universe. Ruskin notwithstanding, a high valuation was always placed on 'smooth minuteness'.[43] Nothing demonstrates this better than Neogothicism, which was close in spirit to Neoclassicism. It served a longing for correctness, for logic, for a universal style and a utilitarian function in architecture; and it was rejected because, to use Matthew Arnold's terms, modern thought continued to be at odds with modern feeling. Modern feeling demanded the disciplines of style and dogma, an art that served society, the beauty of symmetry and wholeness. Modern thought demanded evidence of the free imagination, a society that served art, the beauty of irregularity and ruin. In the last two decades of the nineteenth century the cult of ruins again dominated the response to architecture, and approval for

* and arbitrariness is a lie, a falsification of the antique, a snare set for posterity. The better the restoration the more insidious the lie and the more triumphant the deceit.

restoration was deleted from the catechism of aesthetics. When Maurice Barrès saw barone Giorgio Franchetti's scrupulous reconstruction of the Ca' d'Oro, he regretted the pathos of the mutilated original. 'L'harmonieuse, l'aérienne demeure', he wrote,[44] 'ne demande plus notre compassion, elle prétend à notre hommage admiratif . . . Je me sentis froid pour un art qu'aucun mystère ne baignait plus.'* It was again widely proclaimed that in order to be truly finished a building needed to be in decay. 'C'est malheureux', wrote Proust in 1907, 'que Viollet-le-Duc ait abimé en France en restaurant avec science mais sans flamme, tant d'églises dont les ruines seraient plus touchantes que leur rafistolage archéologique avec des pierres neuves qui ne nous parlent pas et des moulages qui sont identiques à l'original et n'en ont rien gardé.'† In the third chapter of *Sodome et Gomorrhe* Albertine, instructed by the great painter Elstir in the 'inimitable beauté des vieilles pierres', despises the church of Marcouville-l'Orgueilleuse in Normandy because it has been restored. E. M. Forster's heroine Margaret Schlegel, in the second chapter of *Howards End* (1910), is deeply disappointed in Speyer because 'the cathedral had been ruined, absolutely ruined, by restoration; not an inch left of the original structure'. To Giacomo Boni and his friends ruins became something worth preserving in themselves.[45] 'Ogni edificio', they declared,

> ha un periodo di vita che ci è dato prolungare indefinitamente colle nostre cure; ma se ne danno alcuni . . . che dalla particolare natura o struttura materiale sono trascinati ad inevitabile dissoluzione. In questo caso, ed ove occorra per l'uso e valga la pena di perpetuarlo, se ne eriga in altro luogo un fac-simile, ma si conservi la bella ruina.‡

Boni deeply regretted the loss of the shell of the original Fondaco dei Turchi. 'La stessa ruina', he wrote, 'suggeriva l'idea dell'antichità dell'edificio, antichità espressa nella corrosione dei marmi e nel delicato color d'ambra sulla loro pellicola lucidata, prodotti per l'alternarsi del sole e della pioggia.'§

* The harmonious, ethereal residence no longer begs for our compassion, it demands our admiring homage. An art no longer bathed in mystery left me cold.

† It is sad that Viollet-le-Duc should have spoiled so many French churches by restoring them scientifically but coldly. They would have been more touching as ruins than as archaeological patchworks with new stones that say nothing and mouldings that are identical with the originals and yet have retained nothing of them.

‡ Every building has a life-span that can be prolonged indefinitely with due care; but there are some which, because of their special nature and material structure, inevitably decay. When such a building serves a purpose and is worth keeping, let a copy be constructed somewhere else, but let the beautiful ruin be preserved.

§ The ruin evoked the idea of the antiquity of the building; antiquity expressed in the corrosion of the marbles and in the delicate amber colouring and surface polish, produced by the alternation of sunshine and rain.

It would be easy to multiply examples of this fixation with ruins. It was a feature of a mentality whose ancestry was traced by Verlaine to 'L'Empire à la fin de la décadence'. In the late nineteenth and early twentieth centuries literary Europeans identified themselves as 'une race à sa dernière heure'. Transferring their interest from the outer to the inner life, they cultivated 'la psychologie morbide de l'esprit qui a atteint l'Octobre des sensations'.[46] They therefore found special satisfaction in contemplating the mouldy and the rotten; and infatuation with relics of the past, as achievements that could never be surpassed or equalled, was a natural variant on Byzantine postures of exhaustion and sterility. The beginning of the Decadent Movement coincides with the collapse of Neogothicism. It dates from the mid-1870s, when the influential French critic Paul Bourget rediscovered Baudelaire, Swinburne, and the Pre-Raphaelites.[47] It was imported into Britain by Oscar Wilde and George Moore, and into Italy by Gabriele d'Annunzio; and it needs no forcing of the evidence to link the rediscovery of Ruskin to this change in cultural fashion. Proust recognized a proto-Decadent in Ruskin. He discerned an idolatrous worship of beauty behind the professions of allegiance to truth; an affinity with Gustave Moreau and with Ary Renan's principle of 'Belle Inertie' in the denunciations of violence and exertion; a fetishism of symbols in the predilection for the Italian Primitives; and above all a Byzantine sensuality in the hieratic prose. Proust was in St Mark's when he read Ruskin's famous passage on the fall of the Venetian Republic, and he realized that its qualities were those of the great basilica itself.[48] 'Comme l'église byzantine, [elle] avait aussi dans la mosaïque de son style éblouissant dans l'ombre, à côté de ses images, sa citation biblique inscrite auprès.'* Few writers, certainly, were more conscious than Ruskin of being part of a civilization that was depraved and diseased; and few have been so morbidly fascinated by the evils they condemned. Ruskin can at times seem nearer even than Swinburne to Baudelaire; and never more so than when evoking 'l'horreur [qui] tourne aux enchantements' in urban decay:

It is a ghastly ruin; whatever is venerable or sad in its wreck being disguised by attempts to put it to present uses of the basest kind. It has been composed of arcades borne by marble shafts, and walls of brick faced with marble: but the covering stones have been torn away from it like the shroud from a corpse; and its walls, rent into a thousand chasms, are filled and refilled with fresh brickwork, and the seams and hollows are choked with clay and whitewash, oozing and trickling over the marble,—itself blanched into dusty decay by the frosts of centuries. Soft grass and wandering leafage have rooted themselves in the rents, but they are not suffered to grow in their own wild and gentle way, for the place is of a sort inhabited;

* Like the Byzantine church, it too had its biblical text inscribed beside the images in the mosaic of its style, glowing in the gloom.

rotten partitions are nailed across its corridors, and miserable rooms contrived in its western wing; and here and there the weeds are indolently torn down, leaving their haggard fibres to struggle again into unwholesome growth when the spring next stirs them: and thus, in contest between death and life, the unsightly heap is festering to its fall . . .

The garden gate still swung loose to its latch; the garden, blighted utterly into a field of ashes, not even a weed taking root there; the roof torn into shapeless rents; the shutters hanging about the windows in rags of rotten wood; before its gate, the stream which had gladdened it now soaking slowly by, black as ebony, and thick with curdling scum; the bank above it trodden into unctuous, sooty slime: far in front of it, between it and the old hills, the furnaces of the city foaming forth perpetual plague of sulphurous darkness; the volumes of their storm clouds coiling low over a waste of grassless fields, fenced from each other, not by hedges, but by slabs of square stone, like gravestones, riveted together with iron.

The first passage, from the second volume of *The Stones of Venice*, describes the unrestored Fondaco dei Turchi; the second, from *The Two Paths* (1859), a cottage in Rochdale. But more obvious in both than any sense of place is a state of mind. Similar vortices of language, circling the same idea, recur throughout his writing. They bespeak a profound and powerful *nostalgie de la boue*. This obsession drew Ruskin to Turner, whom he celebrated in the fifth volume of *Modern Painters* as the artist of dinginess, dirt, ruin, and litter—the distiller of beauty from urban squalor; and it drew the Decadents to Ruskin, as a harbourer of abnormal predilections and strange perversities. The mad anchorite of Coniston was ripe for rediscovery at the *fin de siècle*, both as a man whose life belonged to the literature of pathology, and as a writer whose work belonged to the pathology of literature.

# · VII ·

## *The World's Inheritance*

AFTER the middle of the nineteenth century historiographic revision absolved Venice from the odium inflicted by Sismondi and Daru. Twenty-five years later changes in the semantics of sensibility were completing its rehabilitation. Now that 'medieval' no longer signified uncouth and depleted, and 'ruined' had acquired new and pleasing overtones, Venetian architecture was a text ripe for retranslation. Furthermore, the labels 'cruel' and, Oriental' had taken on less pejorative meanings. Decadence brought into vogue that esoteric vocabulary of sadism in which cruelty was linked not to hatefulness but to desirability. Arthur Symons, in Venice in 1908, interpreted the dungeons, the chains, and the torture-chambers of the old Republic as something that made Venice not less, but more alluring. He was attracted by an 'odious beauty in these relics of a time when cruelty was a virtue and Casanova was a spy'. The word 'Oriental' had lost its residual associations of menace and monstrosity as the West had conquered the East, both physically and imaginatively. In the second half of the nineteenth century the territorial Orient, from Turkey to Siam, was subject in greater or lesser degree to French and British imperialism. The Orient of Western literature—of Matthew Arnold and Edwin Arnold, of Gobineau, Fitzgerald, and Kipling—was anchored in the picturesque and colonized by Victorians in fancy dress. The Orient of Western music and Western painting was an erotic pastiche. The Orient of Western universities was a model reconstruction that confirmed both the deviance of the modern East and the ability of the West to impose order on its chaos. In popular conception the East had been reduced to wisdom, charm, and mystery; and the banalities of Venetian history appeared transfigured in Oriental light. 'Mere trade', wrote Charles Eliot Norton in 1880, 'became poetic while dealing with the spices of Arabia, the silks of Damascus, the woven stuffs of Persia, the pearls of Ceylon, or the rarer products of the wonderful regions whence travellers like Marco Polo brought back true stories that rivalled the inventions of the Arabian story-tellers.'[1] It is to this innocent East of the Arabian story-tellers that Venice belongs in Proust's *À la recherche du temps perdu*. 'Le soir je sortais seul', the narrator recalls, 'au milieu de la ville enchantée où je me trouvais au milieu de quartiers nouveaux comme un personnage des *Mille et une Nuits*.'*

The Orient of the older lexicography did not completely disappear. It lurks, most notably, in Mann's *Death in Venice*, giving the novella a nightmarish resonance that negates the visionary, exultant note in Nietzsche's anticipation of the death of Socrates. Mann presents the East as the source of mephitic influences that overwhelm austere, rationalistic philosophy with cruelty and lust. When cholera spreads from Asia, it brings anarchy and foulness, and assimilates Venice to subconscious memories of primeval swamp and jungle. E. M. Forster's Orient, likewise, harbours an incubus of monstrosity. It is extraordinary, obscene, and dark with the darkness of the Miltonic Ancient Night. Yet by taking full account of historical circumstance, Forster was able to resurrect this version of the East without implicating Venice. In a world configured by modern communications and modern imperialism, Venice looked less and less Oriental. De Lesseps' canal had created a new frontier on the mental map. It had pushed back the Orient from east of the Adriatic to east of Suez, and it had put Venice on Britain's great imperial highway. The city was now approached as often from Asia as from Europe, and on the passage from India, as Forster's Cyril Fielding discovers, its Orientalism could not but seem diminished. To the eye accustomed to the chaos of the East, Venice had shape, as well as sumptuousness. 'And though', wrote Forster, 'Venice was not Europe, it was part of the Mediterranean harmony.' It contained 'the beauty of form' and it expressed 'the civilization that ha[d] escaped muddle'. Apollo was sovereign still, and Dionysus remained in exile. Yet Mann too depicted a Venice that belonged in a crucial sense to the West. By making Venice the scene of von Aschenbach's encounter with the Angel of Death, he identified the city with Virgil' s Arcadia—the paradise where Death holds his dominion[2]—and he metamorphosed the gondolier into Charon, the Greek ferryman who rows the souls of the dead across the river Styx. This relocation of Venice in the mythology of Greece and Rome was by now a well-established convention. Charon was a familiar figure in Victorian accounts of the city. 'You think of Classical times,' commented the English traveller Anne Buckland, 'of Charon and his ferry boat; of Dis, the city of the dead; and almost expect to see Rhadamanthus standing on the shore to receive you.' 'To this day', wrote Horatio Brown in 1884, 'the passenger across a Venetian ferry lays his

* At evening I used to go alone into the thick of the enchanted city, where in the middle of some new quarter I would feel like a character from *A Thousand and One Nights*.

obol [a small coin] on the gunwale of the gondola, such as Charon's ghostly fares were wont to do.'[3] Henry James (in *The Wings of the Dove*), Gabriele d'Annunzio (in *Il Fuoco*), Hugo von Hofmannsthal (in *The Death of Titian*), and Arthur Symons (in *The Journal of Henry Luxulyan*) all made this transferral. All, anticipating Mann, helped to repatriate Venice by bringing it within the Classical scenario of death.

So the generation that discovered Ruskin was also the generation that discovered Venice and perceived a tragedy in its impending dissolution. In the last quarter of the nineteenth century, in Britain and France especially, the city exerted an ever more potent fascination. 'Venice', wrote Horatio Brown in 1884, 'become[s] the object of a passionate idolatry which admits no other allegiance in the hearts that have known its power.' 'Jamais', wrote Proust in 1906, 'Venise n'a joui auprès des intelligences d'élite d'une faveur aussi spéciale et aussi haute qu'aujourd'hui.'* Guillaume Apollinaire defined it aptly when he called it 'the genitals of Europe' ('le sexe même de l'Europe'), because it troubled the modern consciousness in the manner of an erogenous zone. It was painted by artists without number; it was the subject of endless books; it was a constantly recurring topic in the press; and there was an orgy of sadness among the refined as it succumbed to change and decay. 'Quelle douloureuse symphonie,' exclaimed Fritz von Hohenlohe, 'quelle navrante élégie, un musicien, un poète ne pourrait-il faire de l'immense tristesse de tout ce qui a disparu à Venise, et de tout ce qui y disparaît tous les jours.'†[4]

Venice now became what it has remained: the quintessence of stricken beauty; the archetype of exquisite corruption. The literary tribute spoke simultaneously of pleasure and of pain—of ecstasy, even, and of anguish. Venice was an autumn city; a September rose; a Gioconda of the waters smiling the enigmatic smile of voluptuousness and fatality; a siren singing in cadences of both seduction and despair. 'Est-ce le chant d'une vieille corruptrice,' wondered Maurice Barrès, 'ou d'une vierge sacrifiée? Au matin, parfois, dans Venise j'entendis Iphigénie; mais les rougeurs du soir ramenaient Jézabel.'‡ Barrès's essay on Venice, in his anthology *Amori et Dolori Sacrum* (1903), carries the title 'La Mort de Venise'. In it he hardly mentions the famous sites and monuments. Instead he explores the fetid, cemeterial recesses of the city and the lagoon, obsessed with processes

* Venice has never enjoyed so high and special a favour among the educated élite as it enjoys today.

† What an afflicting symphony, what a heartbreaking elegy, a composer, a poet, could make from the immense sadness of all that has vanished in Venice, and of all that vanishes there every day.

‡ Is it the song of an old corruptress, or of a sacrificial virgin? Sometimes in the morning in Venice I heard Iphigenia; but the crimson colours of evening brought back Jezebel.

of putrefaction that are accomplishing the death not only of the city but of himself: 'A chaque fois que je descends les escaliers de sa gare vers ses gondoles, et dès cette première minute ou sa lagune fraîchit sur mon visage, en vain me suis je prémuni de quinine, je crois sentir en moi qui renaissent des millions de bactéries. . . . On y voit partout les conquêtes de la mort.'*

Death and Venice were now indissolubly linked in the European and American imagination, and anticipating the one became a part of enjoying the other. The pleasure was different from that experienced by earlier generations contemplating the same eventuality. Theirs had been the grim pleasure of grievance satisfied. This was the bitter-sweet pleasure of valediction. It was what Barrès called 'volupté mélancolique', and 'les voluptés de la tristesse'. And precisely because the pleasure was in the anticipation, it became a paramount concern to postpone the death itself. When John Addington Symonds wrote of 'the pathos of a marble city crumbling to its grave in mud and brine', the emphasis was firmly on the present participle.[5] So one of the consequences of the new sensibility was a solicitude for the infirm and injured relic; an angry resistance to any measure likely to turn the prospect of decease into actuality. But at the same time there was resistance to any treatment that would reverse the decline. 'Je plains Venise', wrote Barrès, 'au point où les siècles l'abandonnèrent; mais je ne voudrais pas que ma plainte la relevât.'† The odour of death must be preserved; the process of relentless disintegration sustained. Barrès continued:

> cette agonie prolongée, voilà le charme le plus fort de Venise pour me séduire. . . . Admirons et encourageons ceux qui consolident Venise, mais craignons les 'restaurations', qui sont presque toujours des dévastations. Nous ne voulons pas qu'on paralyse rien, fût-ce une ville morte, fût-ce un ordre d'activité que j'ose appeler la vie d'un cadavre. Il ne faudrait point qu'une discipline générale figeât ces canaux de fièvre et vînt étendre sur la beauté cette perfection convenue qui glace dans les musées.‡

An idea became an obsession and, Faust-like, the Venetophiles forfeited their souls by desiring eternity in a transient enchantment. They wanted a

* Every time I come down the steps of its station towards its gondolas, from that first minute when its lagoon freshens my face, it is in vain that I have taken quinine as a precaution. I seem to feel reborn inside me millions of bacteria. . . . Here one sees everywhere the conquests of death.

† I pity Venice in so far as the centuries have abandoned it; but I would not wish my pity to revive it.

‡ This prolonged agony—there is the Venetian spell that works on me most potently. Let us admire and encourage those who are consolidating Venice, but let us beware of 'restorations', which are almost always devastations. We would not have anything paralysed—be it a dead city, or the sort of activity that I dare to call the life of a corpse. A general discipline must not be allowed to transfix these feverish canals, or cover beauty with the conventional perfection that makes museums so cold.

Venice that was perpetually dying, forever sinking to its grave; and since this was an impossibility, they connived at the manufacture of a fake and consoled themselves with illusion.

In the mid-1870s the British intellectual and artistic élite suddenly decided that, as the heirs of Byron, Turner, and Ruskin, they had a special mission to save Venice from the ravages of restoration and modernization.[6] From this time, measures proposed for the sake of public health, efficient transport, and economic development were regularly reported in the British press and almost always condemned on aesthetic or moral grounds. Venetian conservationists were encouraged and patronized, and the Venetian authorities became a target of constant criticism. They were arraigned as philistine, mercenary, and perversely intent on destroying the pleasures and fantasies of foreign tourists. In January 1876 the *Athenaeum* warned its readers that official vandalism was rife in Venice. The characteristic rosy tint of the city ('which contrasted exquisitely with the green waters and soft grass-green shutters of the windows') was threatened by municipal regulations requiring whitewash; and work about to start on a new road from San Marco to the railway station would 'sweep away several churches and houses of an ancient date and of the highest interest'. The rumours concerning an omnibus road to the station, which also alarmed Robert Browning, seem to have originated with the widening of the Calle San Moise to create the new Via XXII Marzo, leading from the Piazza San Marco to the Ponte delle Ostreghe. *The Times* reported in November 1879 that demolition along this route was creating a street 'of a width quite useless in a town without horses or carriages'. British Venetophiles never forgave the municipality for leasing the small and verdant island of Sant'Elena at the south-eastern tip of the city to the Società Veneta d'Imprese e Costruzioni Pubbliche, which manufactured railway carriages. Robert Browning called it 'a deed of unutterable barbarism'. The *Builder* demanded to know 'whether there was no site but Sant'Elena to found a carriage factory on'; and in a long letter to *The Times* the novelist Ouida inveighed against the factory smoke and mourned the loss of the deserted convent with its rose-covered cloisters. 'There is nothing on earth equal to Venice,' she asserted, 'nothing like it anywhere. Is the greed of a handful of speculators . . . to be allowed to pollute that magical air, those divine skies?'[7]

The introduction of steam transport into Venice set the British press buzzing with indignant voices. After having been brought into public service on the lagoon in 1872, *vaporetti* made their appearance on the Grand Canal in 1880, when the gondoliers went on strike in protest. The *Athenaeum* at once predicted that they would create a turbulence damaging to buildings and quays and make a noise that would end for ever the preternatural Venetian peace. 'It is astounding', declared the journal, 'that a city the municipality of which acknowledges that it depends entirely on tourists and visitors should not only take every occasion to destroy the ancient monuments which are its attraction, but must insist on offending taste by erecting iron bridges and allowing steam launches to ply.' In a letter to the *Builder* in 1883 a sanitary engineer called Hal J. Webber wrote that 'the idea of steamers in Venice [was] about as horrible as an omnibus to Jerusalem or a tunnel penetrating the Channel', and an editorial comment echoed his sentiments. 'It does seem lamentable', said the *Builder*, 'that we cannot preserve some of the unique character of such a place from being destroyed by the all-levelling spirit of modern trade and its servant steam.' Ouida reserved the full force of her anger for these intruders. The *vaporetti*, she insisted, were not only threatening the livelihood of the gondoliers. Their smoke was polluting the atmosphere and their wash was weakening the foundations of the *palazzi* along the Grand Canal. Another article of impeachment against the Venetian municipality was framed by William Stillman, the American journalist who was correspondent of *The Times* in Italy. 'It is', he claimed, 'continually allowing monuments of historical importance and the highest local interest to be taken from housefronts and courts and sold abroad.' Venice in the role of beautiful victim, martyr to modern brutality, was provoking a concern that was ever more strenuous and protective. 'She is the Andromeda of Europe', wrote Francis Marion Crawford in 1905, 'chained fast to her island and trembling in fear of the monster modern progress, whose terrible roar is heard already from the near mainland of Italy, across the protecting water. Will any Perseus come down in time to save her?'[8]

Perseus in fact was already on the scene. He had descended in the late 1870s, in the form of a collective effort to halt the restoration of the basilica of San Marco. This activity, which had been going on virtually unnoticed for thirty years, suddenly triggered an international row in which Venice figured for the first time as an heirloom too precious to be acknowledged as Italy's alone. The force of Anglo-Saxon remonstrance now put Venice high on the agenda of international debate and completed the reversal of its reputation. The monument of human pride, the uncouth provincial city redeemed by nature, disappeared; and in its place emerged a unique and priceless inheritance, universally cherished and adored. Venice, declared *The Times* in 1879, was 'the pride and possession of the whole world'.[9]

Ruskin had called the Ducal Palace the central building of the world. He might equally well have called the tenth-century basilica beside it the central ruin. San Marco, whose richly decorated west front fills most of one side of the great Venetian Piazza, was in the middle of the nineteenth century the quintessence of 'shattered majesty'. It was literally collapsing under the weight of its own magnificence. The marble veneers and columns, the statues, bronze horses, bas reliefs, and friezes that had been brought back from the East and loaded on to its walls after the fourth Crusade, were dragging down the crude brick-and-mortar masonry. Furthermore the structural piers that supported the floor and the five massive cupolas of the roof were sinking into the soft subsoil, causing the tessellated pavement to crack and undulate and the Byzantine mosaics of the vaults and domes to shower down as brightly coloured fragments. San Marco was, in the words of a modern expert, an 'invalid', and it carried the evidence of countless makeshift repairs and haphazard ministrations. Over several centuries its various parts had been patched, propped, buttressed, cemented, and girded; and to some Victorian observers it seemed unlikely that it could stand for much longer. William Stillman remembered that in the early 1860s the southern side of the building 'was in such a state of ruin that only the abundant shoring had prevented the façade from top to bottom from falling bodily into the Piazza.' Even conservationists agreed that the state of this part of the edifice was critical. Zorzi wrote of 'conditions such as to render necessary immediate measures to save it from perishing' ('condizioni tali da rendere necessarie subite pratiche a salvarlo dal deperimento'). Other witnesses described the interior as pitifully deteriorated. The pavement was in loose pieces ('every trampler's foot', said Browning, 'was apt to send flying a lozenge of red or green marble'), and large areas of the decorated vaults were bereft of their mosaics ('tourists carry away the fragments', reported the *Building News*). The programme of restoration carried out by Meduna between 1865 and 1875 was intended to make the building structurally sound, stylistically coherent, and in appearance new. His new foundation for the southern façade, consisting of 2,000 piles overlaid with stone, checked the chronic subsidence at this point and was generally acknowledged as a triumph of engineering. But his success in achieving his other aims ruined his career and made him an official scapegoat. He was delivering Viollet-le-Duc to a public that wanted Ruskin. He removed the anachronistic exterior of the Zeno Chapel in order to restore the 'Byzantine' purity of the building; and instead of making the new work look old, he made the old look new. When he began his operations in 1865 his ideas were still widely accepted as legitimate architectural principles; but when he took down his hoarding and revealed the completed first stage of his enterprise ten years later they were rapidly going out of date. Buildings were now valued less as essays in style than as organisms rooted in national history, and aesthetic preference was shifting from the spick-and-span to the divine patina of age. So the rebuilt southern façade, with its freshly cut veneers, cleaned columns, straightened edges, and rectified inclinations, threw the European élite of taste and intellect into a collective *crise de nerfs*. Henry James reckoned that it was 'a sight to make angels howl'. 'The effect produced', he wrote, 'is that of witnessing a forcible *maquillage* of one's grandmother'. The French historian and architect Charles Yriarte, after examining Meduna's work on both the outside and the inside of the church, complained of

> des pâtes de verre de Murano à la place des pierres rares; des mosaïques neuves au lieu des mosaïques primitives; des colonnes râclées et poncées, dont les fûts jurent avec les chapiteaux; de grands surfaces grises; revêtement vulgaire à veines perpendiculaires, substitué à des marbres précieux du ton le plus riche et le plus harmonieux.*[10]

True to the rigorous lessons of his architectural training, Meduna had compensated both for subsidence and for the shallow concave curve along which the medieval builders had planned the principal façade. Consequently the reconstructed south-west corner was out of line with the west front. In plan it projected 15 centimetres beyond the original work, and in elevation it was 12 centimetres higher. It seemed that Meduna's intention was to rectify the hiatus by demolishing the west front and rebuilding it to align with his new corner. So the old asymmetrical ruin, polished and tinted by time and weather, would disappear; and in its place would be a modern substitute, scraped and smoothed and with lines as straight and angles as accurate as theodolite and chain could make them. The conservationists in Venice reacted with horror and rallied behind Zorzi in his efforts to stop Meduna. In his famous *Osservazioni* Zorzi rhapsodized about 'la divina patina dei secoli' ('the divine patina of the centuries') and demanded the illusion of age as compensation for the loss of antiquity. The paramount concern of restorers, he said, should be to make old the new, not new the old ('far vecchio il nuovo, non nuovo il vecchio'). The old masonry of San Marco should not have been cleaned. Rather the new masonry should have been darkened with soot. ('Dovevasi tingere con la fuliggine il moderno e rispettare il vecchio'). One of Zorzi's chief allies was Pietro Saccardo, a member of the *fabbriceria* (governing body) of San Marco who had tried to halt the demolition of the rear of the Zeno Chapel. In a series of articles in the *Veneto Cattolico* he argued that in the case of San Marco the remedy had proved worse than the disease and that the reconstructed angle should

* Murano glass paste in place of rare stones; new mosaics in place of old ones; scraped and pumiced columns, whose shafts clash with their capitals; broad grey surfaces; common veneering with perpendicular veining substituted for precious marbles of the richest and most harmonious tints.

be demolished and rebuilt in alignment with the main façade. Like Zorzi he lamented the loss of 'il nobile vernice del tempo' ('the noble gloss of time') and suggested that this be imitated on new work with oil and varnish.[11]

Taste was already changing and it was not difficult to persuade both the Venetian public and the Roman government that Meduna's methods were barbaric. Popular attitudes followed educated opinion, and almost overnight approval changed into hostility. Preference settled in favour of a new style of treatment for ancient buildings. It was called conservation and it was promoted in the name of Ruskin. However, the conservationists' addiction to the aesthetic of beautiful decay allowed few concessions to Ruskin's lamp of truth. Ruskin had decreed that repairs should never be disguised. The new should look new and the observer be left undeceived. But in Venice conservation pandered to deception and made the new look old. Even Meduna, desperate to retrieve official and popular favour, was apparently prepared to fumigate his masonry in an attempt to simulate the effects of age. He denied, furthermore, that it had ever been his intention to demolish the west front. But Meduna had little further influence. In May 1879 the Ministry of Public Instruction in Rome, alerted by the local agitation, ordered all work on the basilica to be suspended pending an inquiry by the Commission for the Preservation of Ancient Monuments. Within a year the Commission had formally recommended a policy of conservation, and the minister ordered that the west front be repaired rather than rebuilt. A code of regulations was published and a *Commissione di Vigilanza* set up. Saccardo was appointed Meduna's coadjutor and then, after Meduna's death in 1883, his successor as director of operations.[12]

It was during this bureaucratic interlude that the affair became a European *cause célèbre*. The German press began to express alarm at what was happening in Venice; in Paris a Comité des amis de Saint Marc was founded; and in Britain the Society for the Protection of Ancient Buildings organized a campaign to prevent further mutilation of the basilica. The British effort was loud and assertive, but it was too late and seemingly too Italophobic to have much influence in either official or artistic circles. In Rome it was asked why the British had never protested against the Austrian restorations. In Venice it was noted that the battle was all but won before the British joined in, and it was feared that their interference would antagonize more people than it converted. Boito published a barbed, sarcastic article in the *Nuova Antologia* which revealed both his own unhappiness with the treatment that the basilica had received and his exasperation with the style and the timing of the British intervention. 'La più acerba invece di tutte quante le accuse', he complained, 'è quella che da una nazione straniera capita con grandi parole rettoriche, e con chiassose e insistenti e affettate manifestazioni pubbliche.'* The gist of his article, which was designed to soothe wounded amour-propre, was that the Italians had nothing to learn from foreigners—least of all from the British, whose treatment of their own ancient monuments left much to be desired. The Italians were their own best critics, and had long since outgrown the mistakes that the British were only now discovering.

Zorzi's pamphlet had not in fact gone entirely unnoticed in Britain at the time of its publication. The *Athenaeum* had published an account of the work in November 1877. But it had tried in vain to stir up influential opinion. 'Why', it demanded, 'has not Mr Ruskin, or Mr Browning, or Mr Rossetti, or Mr Alma Tadema, or Mr Wallis, or Mr Burne-Jones, or Mr Holman Hunt delivered his soul by protesting against this astounding Philistinism?' The reason was that Ruskin, beset by mental illness, was now as unresponsive and innocuous as a wasp in October, while the others were content to wait for an initiative from the SPAB. This was delayed by the indifference of William Morris. During his first visit to Venice, in the spring of 1878, Morris found himself unmoved by St Mark's and uninterested in the controversy. 'Even the inside of St Mark's', he wrote to Georgina Burne-Jones, 'gave me rather deep satisfaction, and rest for the eyes, than that strange exultation of spirits which I remember of old in France.' And he added: 'I don't think this is wholly because I am grown older, but because I really have had more sympathy with the North from the first. . . . Even in these magnificent and wonderful old towns [of Italy], I long rather for the heap of grey stones with a grey roof that we call a house north-away.' It was not until the end of 1879, fully two and a half years after Zorzi's pamphlet had been published, that he brought the SPAB into action. By now he realized that this was a way to boost interest in the Society. He told Robert Browning that the restorers of St Mark's 'have given us an opportunity of appealing to people who might not otherwise be easy to move'. The SPAB then made up for its lateness by exploiting to the full all the resources of the Victorian campaigning tradition—letters to the press, public meetings, scientific investigations and reports, and a memorial with 2,000 signatures, some of them very eminent. This last item it addressed to the Italian government.

For six months, throughout the winter and spring of 1879/80, it kept people talking and thinking about St Mark's; and then, in the summer of

* however, the most bitter accusation of the lot is that which is levelled by a foreign nation with big rhetorical words, and with clamorous, peremptory, and affected public demonstrations.

1880, when the interest of the British public began to wane, it set up a separate St Mark's Committee to promote and co-ordinate an international response. The secretary was the Pre-Raphaelite painter Henry Wallis, and his principal correspondents abroad were Charles Yriarte in Paris and Charles Eliot Norton in Massachusetts. Yriarte was especially energetic in drumming up support. He publicized the cause in the French press and recruited a constellation of notables for the Committee, including the connoisseur baron Adolphe de Rothschild; the director of the prestigious *Revue dex Deux Mondes* Charles Buloz; the architect Charles Garnier; the artist Jean Meissonier; and the historians and critics Paul de Saint-Victor, Edouard Charton, Louis Gonse, Charles Blanc, Paul Mantz, and Eugène Veron. Zorzi was invited to join the Committee and accepted 'con vivo piacere'—but he was diffident about approaching his compatriots. 'Je crains de ne pouvoir faire de prosélites', he told Wallis in rather uncertain French. 'J'ai beaucoup d'ennemis pour avoir soulevé et soutenu, le tout premier, le [sic] cause de Saint Marc.' And he was cautious about canvassing the Italian press. 'Quant aux journeaux [*sic*] il faut s'abstenir encor [*sic*] jusqu'à ce que les membres italiens de notre société seront plus nombreux, d'autant plus que dans la question de Saint Marc et d'autres monuments de Venise ils sont actuellement presque tous d'un avis contraire au mien.'* Zorzi, quite plainly, was hedging. Like Boito, he feared counterproductive effects from the efforts of foreigners whose facts were not always straight and whose accusations were out of date. In August 1880 he wrote to correct several errors in the draft of the Committee's circular and pointed out that San Marco was no longer in danger. 'Riguardo ai progetti per i lavori ulteriori da farsi in San Marco, sembra che si voglia adottare un altro sistema, un sistema migliore . . .'† Zorzi was not alone in his apprehension. Norton reported from America that Longfellow declined to join the Committee, 'on the ground that . . . anything that looks like foreign interference will do more harm than good with the sensitive Italians'. Turgenev, on the other hand, wrote (in English) from Bougival agreeing to join and asking how he could best promote its aims in his native Russia. This reply was typical. Letters received by Wallis were convincing evidence that Venice had become an object of international concern. When finally published the manifesto of the St Mark's Committee listed well over a hundred members,

* I fear that I shall be unable to make converts. I have many enemies as a consequence of being the first to raise and plead the cause of St Mark's. . . . As for the newspapers, we should hold off until the Italian members of our society are more numerous, especially since almost all the papers are currently of an opinion opposite to mine in the matter of St Mark's and other Venetian monuments.

† Concerning plans for further work to be carried out on St Mark's, it seems that they want to adopt a different system, a better one.

representing eminent scholarly and cultivated opinion in Britain, France, the United States, Germany, Austria, Switzerland, Holland, Belgium, Poland, and Russia.[13]

It is an indication of the strength of feeling in Britain that the campaign was not more widely ridiculed. The dean of Christ Church Oxford, forgetting why Byron had been refused a memorial in Westminster Abbey, proclaimed that Venice had been made sacred by Byron's poetry. George Edmund Street, impeached by the SPAB for his own ruthless restorations, denounced vandalism in Venice and flourished a piece of mosaic from the baptistery of St Mark's ('purchased of the sacristan for a few lire') to make his point. This sort of crankiness encouraged the rearguard of the Gothic Revival to scoff and mock. The *Building News* detected sensationalism in the reports of the 'ladies, antiquarians, and amateurs' of the SPAB, and the *Builder* likewise was suspicious of its 'ill-regulated enthusiasms'. William Stillman was convinced, after personal investigation in Venice, that Meduna had been libelled and his intentions traduced. Judged from an architectural standpoint the indictment was unjust and the alarm needless. The fuss was the work of a few artists who were making a fetish of 'picturesque tints and pictorial decay'. But these voices, muted and uncertain, were echoes from the past. Public and professional support for the campaign at home and abroad showed clearly that the attitudes to Venice and to restoration nurtured by Gothic Revivalists were superseded. The counsels of European art and literature were dominated by a new mentality—that of the Venetophile, who claimed Venice for the world, safeguarded it for the future, and loved it above all for its picturesque tints and pictorial decay. *The Times* diagnosed the agitation in Britain as 'a pleasing testimony of the sincere feelings of Englishmen towards art and archaeology', and it judged that since St Mark's 'belong[ed] from of old to art, not to Italy alone', it was right that lovers of art should act to protect it from 'the dangerous stage of a restoring temper such as smoothed out the venerable wrinkles of many an English church and mansion twenty or thirty years ago'. The *Architect*, likewise, warned the Italians against the mode of restoration 'encouraged so fatally . . . until quite recently' in Britain. 'The only proper line of argument to take', it went on, '. . . is to expound again and again the principle of preserving the *status quo* of the truly interesting antique, upon the simple ground that posterity has the same proprietary interest in it that we have.'[14]

The *leitmotiv* of the campaigners' propaganda was the magic of time's artistry. 'Those who [saw] St Mark's twenty years ago', wrote Henry Wallis, 'remember the patina which the marble surfaces had acquired from time, and which time only can bestow.' Street's opinion was that the true value of

St Mark's derived not from its architecture but from its colour—'the lovely colour given by centuries of exposure'. Charles Eliot Norton enthused about 'this unique façade, to which the hand of time has given the last touch of beauty in the hue which only years can bestow'; Charles Yriarte, writing of the stones of San Marco, rhapsodized about 'la riche patine dont le temps les a revêtus'; and the memorial sent by the SPAB to the government in Rome described how time had glorified the original fabric of the basilica. 'It has cast a veil of beautiful tone over the surface, which no device of man's hand could accomplish; it has softened whatever was crude, without hiding anything that was delicate; it has, we may say, restored those rare and laboured stones to nature without taking them from art.' The truth is that the Venetophiles wanted the finish bestowed by time without the infirmity created by age. So although they were steeped in Ruskin's rhetoric they were impervious to his morality. Ruskin had demanded honest acknowledgement of a building's mortality; but they refused to contemplate the possibility that the basilica might soon be nothing but a memory. 'If St Mark's be ruinous', wrote the architect John Pollard Seddon, 'it should be restored as it is. It cannot be allowed to crumble into dust, as some over-zealous but rather selfish of its admirers would wish.' 'We are confident', wrote the SPAB memorialists, 'that if it be threatened, it is within the power of science to devise a remedy which would restore its stability without moving a stone or altering the present surface in the least.'[15]

Pietro Saccardo, who agreed to become a member of the St Mark's Committee in October 1880, gave the Venetophiles what they wanted. When his scaffolding and hoarding were taken down in 1886, everything looked more or less as it had done in 1865. The rear altar of the Cappella Zeno was again protruding from the southern façade; the old interior mosaics had been replaced; and the hiatus between the south-west corner and the west front had disappeared. Quite how this last result was achieved remained a mystery. The official regulations had directed that Meduna's work be demolished and the angle rebuilt; but some observers suspected that Saccardo had, in fact, carried on with the rebuilding of the main façade. According to Stillman, Saccardo had 'completed' what Meduna had begun; and the opinion of the *Builder* was that the whole of the west front had been renewed. Two inspectors from the SPAB reported that Saccardo's work at the south-west angle was indistinguishable from Meduna's. Substance was added to supposition by a pamphlet published by the *fabbriceria* of San Marco in 1883. This reviewed recent work on the church and announced that half the main façade had been dismantled and reconstructed without anybody having noticed. ('In questi ultimi anni metà della facciata principale fu letteralmente disfatta e ricomposta . . . senza che nessuno ci potesse trovare innovazione di sorte'.) Saccardo himself admitted that he had heightened a portion of the west front, though he never explained how and challenged the public to discover exactly what had been done. He claimed that he had merely raised what Meduna had lowered—which started yet more speculation about what Meduna had been up to. But the truth of the matter was no longer important, because the illusion of time's ruin remained intact. Saccardo's achievement was as unobtrusive as his pride in it was loud, and even the most alert conservationists were mollified. 'I won't say that they do as much harm as they used to do,' Boni told William Caroë, 'for they do not scrape, they do not rebuild entirely.' Boni and his friends were embarrassed by Ouida's letter to *The Times* of September 1885 because it accused the ecclesiastical authorities of practices that had by then been abandoned. He told Philip Webb that the mosaics were no longer renewed and that 'the principle of the preservation of monuments in Italy now [was] right'. In 1886 the SPAB heard of plans to level the undulating pavement of St Mark's, and sounded another alarm; but Saccardo issued a soothing reassurance. 'The worthy Society', he wrote, 'should know my ultra-conservative ideas in relation to monuments, and especially in reference to the basilica of St Mark's, inasmuch as they have for some years prevailed and become known. . . . No one thinks of correcting this peculiarity of the basilica.' He was as good as his word. Boni reported to Webb in August 1888 that the ruined and sunken areas of the pavement were being reset, 'without aiming however to obtain a level surface, and restoring the original tesserae on the same spot which they occupied'.[16]

By the end of the century, in Venice as elsewhere, conservation was the new orthodoxy. When people spoke of restoration they now meant making the new look old, not making the old look new. The Gothic Revival was regarded as a superseded barbarism. 'How changed are the times!' observed the *Builder* in 1902. 'Public opinion has completely gone over to the conservative side of the question; the "Restorers" are nowhere.' Discussing San Marco, it contemplated with horror 'the atrocious restorations of twenty to thirty years ago . . . and the machine-made "Gothic art" of the period, with its crude colouring and lifeless drawing'. Modern practice, it noted with relief, conformed to fundamentally different principles. 'Now the object in view is to produce an untouched appearance.' Deception had become the restorer's aim. During the twentieth century major structural surgery was frequently needed to keep the moribund building standing. Yet each time the hoarding and scaffolding were taken down, nothing was seen to have changed.[17]

These skills of imitation and repair, ever more sophisticated and ever more widely applied, enabled old Venice to survive. However, it survived in idea, not in substance. There was never a funeral; but the organism perished none the less. Behind the screens and canvases that became a permanent feature of the Venetian scene, a metamorphosis was in operation whose ultimate outcome must be a city that was not dying, but dead with the deadness that had never lived. Full of medieval masonry that no medieval hand had ever touched, resplendent with Renaissance decoration that no Renaissance eye had ever seen, Venice was acquiring an immortality that recalled, to use William Blake's expression, 'something other than human life'.

The Ducal Palace was another of the many buildings that began to turn into imperishable effigies. Like San Marco, it was partly hidden by hoarding and builders' paraphernalia throughout the later 1870s and the 1880s. Between 1876 and 1889 it underwent treatment that was every bit as drastic as that endured by St Mark's; but since the repairs were skilfully disguised, they were far less controversial. Ruskin had noted the perilous state of the building in 1852. He had predicted that it would not survive another five years, since the massive capitals of the lower arcade, lovingly and exhaustively described in *The Stones of Venice*, were 'so rent and worn'.[18] One of them at least was later discovered to have fragmented into more than thirty pieces. The damage had been caused by oxidation of the iron tie-rods; by the unequal thickness of lead linings between the capitals and the shafts; and by the poor quality of much of the original workmanship. (Some of the mortar was found to be unmixed.) The shafts themselves were also fractured and corroded—as a result of frost and fire in the lower arcade, and of uneven lead linings below the bases in the upper arcade. Everywhere on the two principal façades columns were leaning and voussoirs were slipping. The south-east corner (the Vine-Tree Angle) was subsiding, while the south-west corner (the Fig-Tree Angle) was in danger of collapsing under the enormous weight of the solid wall above. During the work of restoration the foundations of the two façades were strengthened; the five arches on the south that had been bricked up since the fire of 1577 were opened; and the entire south-east corner, adjacent to the Ponte della Paglia, was demolished and rebuilt. Most of the damaged capitals were repaired with cement and copper bonding; but 13 out of the 37 on the lower storey, and 27 of the 73 on the upper level, were found to be irreparable and were replaced by copies. Each replica occupied a sculptor for a period of eighteen months to two years and was carefully stained and blackened to give it the appearance of age.[19] Boni, who took part in this work as assistant to the chief architect, Domenico Rupolo, regretted the loss of so much original fabric. 'The old age of such buildings', he told Webb, 'is their age of beauty.' He was adamant that much had been discarded that could have been saved, and that the copying was often inaccurate and coarse. Saccardo was of the same opinion. In 1899 he published in the *Nuovo Archivio Veneto* a paper pleading for the old capitals, now exhibited as museum specimens inside the Palace, to be reinstated.[20] But Rupolo's assiduity in making the new look old disarmed criticism, and this time the consensus was one of approval. The *Builder*, noting that the famous column of the south-west angle had been 'replaced and coloured with so much adroitness that it requir[ed] very close examination to detect it', called the work 'a marvel of skilful restoration'. William Stillman thought that very few people would be able to tell which of the capitals were new. 'The stains, the marks of time and weather,' he assured readers of *The Times*, 'have been so perfectly imitated on the new stone, that the closest scrutiny is necessary to see what is weather-worn and what is artificially treated.' Horatio Brown agreed. 'The uneducated eye at all events', he wrote, 'cannot distinguish between the old and the new.' Even Ruskin, in the more benign interludes of his old age, gave his blessing. Enid Layard recorded his saying that the restorations were 'very well done and one could not detect which of the capitals were new'.[21]

Death had become an essential ingredient in the appeal of Venice; but the death envisaged by the Venetophiles was a slow and ravishing obliteration whose consummation was always in the future. Such a death could be experienced imaginatively but could never occur in actuality. It was in the nature of their infatuation to discount the possibility of death as an imminent and sudden end. This became apparent during the debate about St Mark's, when three British architects (Street, John Stevenson, and William Scott) refused to acknowledge that the west front of the basilica was structurally unsound. They insisted that it had no need of repair. In fact it was part of the paradox of the concept of Venice as a dying city that the stability of its buildings had become legendary. 'These structures never fall—never', proclaimed Edmund Flagg in 1853. 'Such a thing, common enough . . . in . . . the New World on *terra firma*, is never heard of in old and sea-girt Venice . . . The edifices of Venice, even the humblest, seem constructed for eternity, not time.' The thousand-year-old Campanile in the Piazza San Marco, the most prominent landmark in the Adriatic, was often cited as an example of this durability. Its strength was proverbial. 'Incrollabile come il Campanile di San Marco' was a popular local saying.

The long survival of this tower in defiance of earthquake, lightning, fire, and all the injuries of mundane use and abuse, seemed a rebuke to anyone who doubted the permanence of Venetian architecture. S. B. Burton of Newcastle, who introduced himself as 'a working mason, foreman, and builder', wrote to the *Architect* in 1880 to point out that the solidity of the belltower made it inherently improbable that the basilica was unstable. 'When we look at the sound condition of the massive Campanile, but a few yards removed and more than three hundred feet high . . . we cannot for a moment admit the supposition without very careful examination.' Horatio Brown confirmed that 'it seemed to almost every expert, and certainly to all the profane, the soundest building in Venice'.[22]

Widespread shock and disbelief therefore followed the events of 14 July 1902. Shortly after half-past nine on the morning of that day, when the gilt angel at the summit of the Campanile was glittering in the summer sun, clouds of lime-powder suddenly spurted from the tower about 6 metres from the ground. Huge cracks then ran up the curtain walls between the piers, the base of the building bulged and split apart, and the whole structure collapsed into a pile of 18,000 tons of dust and rubble. There was little concussion, but the atmospheric disturbance almost capsized a steamer on the lagoon. 'From a distance I saw the angel slowly descend, swaying, but upright', recalled an American eye-witness. '[It] descended in an upright position until the cloud of dust rose and covered it, and it must have come down a full 100 feet before toppling. When found it was directly under and within the main portal of St Mark's.' All over the surrounding area a fine snow-like deposit settled on pavements, parapets, sills, and ledges. Gas escaping from ruptured pipes filled the air with the acrid stench of onions. By a miracle there were no human casualties, and the only other building to suffer damage was Sansovino's library, which had a gash torn in its side. Nevertheless, the European press talked of a major catastrophe. 'Venice will never be quite the same again,' pronounced *The Times*. 'So glorious a survival of the great age of Venice, so conspicuous a point in the landscape, and a building glorified by such a series of great painters from Gentile Bellini to Canaletto, from Guardi to Turner, belonged to the whole world and the whole world will mourn its disappearance.' The *Builder* confessed to 'a sense of incredulity, of hardly believing our own eyes'. And it agreed that 'one cannot feel that Venice can ever be the same without it'.[23]

Illusions were shattered; but they were soon repaired. Within hours of the disaster, the *sindaco* of Venice, Count Filippo Grimani, announced that the Campanile would be rebuilt as it had been and where it had been ('come era e dove era'). An appeal was launched for funds; the remains of the old monument were sifted for reusable material; and an operation was begun to disguise the great erasure. At first there was some demurring. *The Times* thought it would be folly to build twice on ground that was apparently infirm. The *Builder* said that any attempt at reconstruction would be a mockery. 'The Campanile recreated on its old lines must . . . really be a new building, pretending to be the old one, and thus adding one more to the list of architectural shams born of the idea of "restoration". It can never have the archaeological and historic value of the original.' Some people preferred the Piazza without it—either because, like Wilfred Scawen Blunt, they thought belltowers had no meaning in modern life; or because, like Roger Fry, Henry Wallis, and Lonsdale and Laura Ragg, they found that the Piazza was aesthetically more satisfying bereft of its enormous bulk. 'I don't believe', wrote Roger Fry, 'that any artist would have thought of putting one there now that the Piazza has taken its present shape.' The counter-argument was that the Campanile was needed in order to redress the architectural confusion of the square. 'It could never have been contemplated', wrote Colonel Hugh Douglas, a resident of many years, 'that any one, entering the Piazza at the other end, should see at once both the church of San Marco and the Ducal Palace—two buildings which are most discordant and quite impossible when seen together.' But the decision to rebuild was not the outcome of dispassionate debate. It was a concession to popular sentiment, which was just as exigent abroad as it was at home. 'To think of Venice without the Campanile', wailed Edward Hutton, 'is to us all almost an impossibility. It was not the Piazza alone that the famous belltower dominated, but all Venice too, across whose silent ways that bell . . . seemed like an assurance of safety, of our civilisation, of Europe, of our Faith.' The Royal Academy raised £1,000 in response to the appeal and the president, Sir Edward Poynter, said that the amount would have been greater, but 'times [had] not been very favourable to artists lately'. Sustained by all this interest and emotion, the best that Italy could provide of antiquarian ingenuity and engineering skill was set to retrieve the calamity. The thousands of fragments of the shattered *loggetta* of Sansovino were sorted and pieced together like the shards of a precious vase; bricks for the new tower were fired in special wood-burning kilns in an attempt to reproduce the old colour; and massive foundations were laid to ensure that the life of the replica would be at least as long as that of the original.[24]

In the language of politics, the desire to imitate what had perished was translated as a symptom of revival. The facsimile of the Campanile, begun in 1906 and completed six years later, was treated by official Italy as the enterprise of a reborn and resurgent nation. The ceremony of inauguration in April 1912 was turned into a major state occasion. Army, Church, Crown, and foreign dignitaries were ostentatiously in evidence, and the world multiplied its congratulations. The *sindaco* of Venice, Count Grimani, and the Italian minister for public instruction, Signor Credaro, both referred to the tower as an expression of the patriotic spirit that was currently winning national glory in Libya. At a municipal banquet the British ambassador to Rome, Sir James Rennell Rodd, hailed it as a phallic emblem. It was, he said, 'a virile symbol of the good blood that belies itself not'. In a voice trembling with emotion he thanked Venice, which had proved itself 'now and ever worthy of the new Italy'.[25] In architectural circles there was more awareness of a loss of will—or ability—to create something new. 'In any previous period', commented the *Builder*, 'if an important building fell down, the immediate desire would have been not to "restore" it, but to erect something better and finer in its place.'[26] The purpose of the replica was in fact not to register actuality but to perpetuate a fiction—the fiction that old Venice, the beautiful ruin, had been preserved; preserved in order to go on dying. No one could say how closely the reconstructed tower resembled the original. The building had had to be reconstructed from photographs, since no plans or designs or architectural drawings of it existed. There was no record even of its exact height. But after ten years it was difficult to recall precisely the details of the vanished structure. 'I cannot remember the old Campanile with enough vividness to be sure,' wrote E. V. Lucas, 'but my impression is that its brick was a mellower tint than that of the new—nearer the richness of S. Giorgio Maggiore's, across the water.' Maurice Barrès too thought the colour unauthentic. 'Son aspect de neuf', he commented, 'lui donne l'air d'un intrus, l'air d'un géant qui serait venu de l'étranger demander en mariage la basilique et demeurerait là gauche et figé, en costume trop neuf.'* He would have preferred the work of imitation less inhibited by scruples about fakes, and regretted that the brickwork had not been tinted in order to make it look old.[27] However, there was a general willingness to believe that the facsimile was exact. 'When I saw it', said the traveller Cecil Torr, 'I felt that the old Campanile had come to life again.'[28] And that is how it features in the modern guidebooks. So, with the aid of a little goodwill, it fulfilled its purpose. It made it possible for contemporaries to forget, and for subsequent generations not to know, that the great ca-

* Its look of newness makes it seem like an intruder, like a giant come from abroad to request the hand of the basilica in marriage, and who stands there awkward and transfixed, in a suit that is too new.

lamity had ever happened. And the world, secure in its illusion that Venice could, after all, be the same again, went on talking and writing about 'the dying city', 'the sinking jewel of the Adriatic', 'Venice in peril'.

The most scathing attack on the rebuilt tower came from outside Venice, but inside Italy. The group of young Italian writers, artists, and architects who called themselves Futurists denounced the structure and the whole ethos that had produced it. Filippo Tommaso Marinetti, poet, novelist, and chief propagandist of these *enfants terribles*, inveighed (in London, and in French) against 'ces passéistes vénitiens, qui ont voulu reconstruire cet absurde campanile de San Marco, comme s'il agissait d'offrir à une fillette qui a perdu sa grand'mère une poupée en carton et en étoffe qui ressemble à la défunte'.*[29] This was one of many speeches, proclamations, and manifestos in which the Futurists declaimed against all those aspects of old cities that the literature and art of the *fin de siècle* had celebrated.[30] Marinetti abhorred 'la corrosione, il logorio, le sudicie tracce degli anni, lo sgretolarsi delle rovine, la muffa, il sapore della putrefazione'; the painter Carlo Carrà repudiated 'tutti i colori in sordina, anche quelli ottenuti direttamente senza il sussidio truchistico delle pàtine e delle velature.'† In July 1910 these two, with the painters Boccioni and Russolo, carried their campaign into the heart of enemy territory by scattering thousands of copies of a diatribe called *Contro Venezia passatista* from the clock tower in the Piazza San Marco. Tourists returning from the Lido in the hot summer afternoon were bombarded with a declaration of war against the city they adored. 'Noi', bellowed the Futurists, 'vogliamo preparare la nascita di una Venezia industriale e militare che possa dominare il mare adriatico, gran lago italiano.'‡ To this end, they demanded the filling in of the canals with the masonry of demolished *palazzi*; the burning of the gondolas, those 'rocking chairs for cretins' ('poltrone a dondolo per cretini'); and the raising to the sky of metal bridges and of factories crowned with smoke. In the Fenice theatre Marinetti harangued a packed auditorium with the same message, raising a chorus of catcalls and whistles by exulting over the prospect of old Venice drowned and obliterated.[31] When the furious sirocco wind drives the waters of the Adriatic over the helpless city, 'Oh,' he

* those Venetian fogeys, who have set about rebuilding that absurd Campanile of St Mark's just as if they were offering to a little girl a rag doll resembling the grandmother she has lost.

† corrosion, dilapidation, the stains of time, the crumbling of ruins, mould, the smell of putrefaction . . . all muted colours, even those obtained naturally, without the assisting trickery of coating and glazing.

‡ We want to prepare for the birth of an industrial and military Venice, with power to dominate the Adriatic, the great Italian lake.

declaimed, 'come balleremo! Oh, come plaudiremo alle lagune, per incitarle alla distruzione! E che immenso ballo tondo danzeremo in giro all' illustre ruina!'*

This Venetophobia was part of a sustained and stridulous protest against the elegaic note. It was an attempt to smother the refrain of 'far away', 'never more', 'long ago'. The Futurists had had enough of ruins, nostalgia, moonlight, and dreamy distances. They were sick of museums and old masters. Nothing, they decided, could grow in the still and stifling atmosphere of Romanticism, Symbolism, and Decadence. Libraries and academies were fit only for embalmers and taxidermists. To safeguard the creative urge, the young must be rescued from their enervating tutelage to Baudelaire, Poe, Mallarmé, Verlaine, and Ruskin. 'Ce déplorable Ruskin' was especially pernicious. He typified the shrivelled, impoverished sensibility that was incapable of confronting the modern world and the modern experience. 'Avec sa haine de la machine, de la vapeur, et de l'électricité,' Marinetti told the Lyceum Club in London, 'ce maniaque de simplicité antique ressemble à un homme qui, après avoir atteint sa pleine maturité corporelle, voudrait encore dormir dans son berceau.'† The British fixation with Italy's past was attributed to his mummifying influence. And if Ruskin was the supreme pontif of *il passatismo*, Venice, the city with which he was most closely associated, was its *cloaca maxima*. It had become the temple of the cult that the Futurists most abhorred—'l'adorazione della morte'. This was anathema, because it signified morbid enslavement to intellect. Death was the ultimate rationalization ('suprema definizione dell'intelligenza logica'). So the Futurists abolished death by decreeing the end of the hypertrophied intellect and prophesying man the intuiting mechanism—a new Prometheus, rendered indestructible by the constant replacement of his component parts. 'L'uomo meccanico dalle parte cambiabili' would deliver humanity from the idea of death, and hence from death itself. And with death would vanish all desire to arrest the fleeting moment. 'Alla concezione dell'imperituro e dell'immortale,' wrote Marinetti, 'noi opponiamo . . . quella del divenire, del perituro, del transitorio, e dell'effimero.'‡ Modern technology, the apotheosis of metal and speed, was worshipped as a paradigm of timeless, omnidimensional energy—the *élan vital* of Bergson. The machine was to take the place of the nude as the

* Oh how we shall dance! Oh how we shall cheer the lagoon, to incite it to destruction! And what a dance of triumph we'll perform around the illustrious ruin!

† With his hatred of machinery, of steam, and of electricity, this fanatic for antique simplicity is like a fully grown man who wants to go on sleeping in his cradle.

‡ Against the concept of the imperishable and immortal we set up that of becoming, of the perishable, the transitory, and the ephemeral.

proper subject of art, and recollections in tranquillity were to be shattered by the madness of impulse.[32]

Futurism was in many ways characteristic of its time. European artistic life in the pre-war decade was remarkable for the prominence of an avant-garde, who set out to challenge 'culture' and its associated traditions and institutions. In Paris, London, Barcelona, Munich, and Moscow dozens of young painters and critics were, like the Futurists, identifying themselves as 'studenti ribelli di questo mondo troppo saggio' ('student rebels of a world too wise').[33] Inspired by the intuitionist philosophy of Bergson, in which time features qualitatively as duration, and change features organically as flux and continuum, they were labouring to rid art of all the intellectual accretions with which a dominant class had encumbered it. They rejected Cartesian notions of time and space, which—according to Bergson—distort reality; and in their thirst for freedom from bourgeois values they looked for models to primitive and vernacular idioms. Wherever they exhibited, decorum fled, controversy raged, and art became associated with dissent and subversion.[34]

Yet in other, important respects the Futurists were marginal. If people still visited the Louvre and venerated Venice, this was not because the avant-garde failed to educate public opinion. On the contrary, as a result of its work the world was never to look the same again. The old icons of culture retained their allure partly at least because Futurism failed to convince the avant-garde. Marinetti and his gang were hijackers—highly publicized intruders who were trying to force the modern movement to take a direction it did not want to follow. Their art of contemporary images and frenzied displacement had little in common with the modern painting of Paris. In Paris dynamism meant either the dynamism of volume or the dynamism of light. It did not mean the dynamism of mechanical movement. In fact the leading figures of the avant-garde ignored the machine and remained in thrall to nature—to nature as perceived by Cézanne, what is more. Form and equilibrium, *gravitas* not *levitas*, were their dominant preoccupations. Matisse dreamt of an art of balance and tranquillity. Braque and Picasso developed a variety of cubism which the French theorists qualified as Bergsonian but in which the Futurists could see nothing but traditional stasis and inertia. Likewise, the patriotism of the Futurists went against the grain of the avant-garde mentality, which was internationalist and anti-political. The Futurists were rightly accused of being concerned more about the state of Italy than about the state of art. They were trying to revive the aborted Risorgimento, by reclaiming their country from the humiliating ascendancy of culture's cosmopolites. They cared above all about national dignity and national prestige; and they loathed the artistic heritage because it encouraged Italians to prostitute themselves to foreigners. 'Voi volete prostarvi divanti a tutti i forestieri,' Marinetti chided the Venetians, 'e siete di una servilità ripugnante! . . . Siete divenuti camerieri d'albergo, ciceroni, lenoni, antiquarii, frodatori, fabbricanti di vecchi quadri, pittori plagiari e copisti.'*[35]

Consequently Futurism won few converts outside Italy, and those few soon rejected the Italian version for formulations of their own. Opinion even among free-thinking intellectuals was often hostile. Gide called Marinetti 'a rich and conceited fool' ('un sot, très riche et très fat'). Kandinsky rebuked the Futurists for their flippancy. The critics who led Anglo-Saxon opinion in a new direction—Roger Fry and Clive Bell—excluded them from the pantheon of Post-Impressionism. These writers were concerned more to re-interpret and rehabilitate the past than to dance on its grave. Even at their most provocative they never echoed the vandalistic language of the Italians. Clive Bell's sweeping dismissal of post-Byzantine European art was tame in comparison with Marinetti's call for the demolition of museums and libraries and his vilification of Rome, Florence, and Venice as 'les trois plaies purulentes de notre péninsule' ('the three festering sores of our peninsula'). And it is difficult to think of anything further removed from the Bloomsbury creed of 'goodwill, plus culture and intelligence', than Marinetti's aggressive chauvinism and his panegyrics on machinery and war. The Anglo-Saxons who did not ignore the Futurists reacted strongly against them; so, far from damping Venetophilia, Futurism inflamed it. The verbal violence made the cherished city seem even more threatened, and Venetophiles responded with apocalyptic warnings of desecration. 'Great chimneys', cried Edward Hutton, 'will take the place of leaning campanili, vast factories will occupy the foundations of the magical palaces, and a huge industrial capital and port, shrouded in smoke, clanging with machinery, filthy with mud, and groaning with misery, will rise where for so long Venice had her inviolate throne.'[36] The reasons that made Futurism ineffectual abroad might have made it potent at home, but here Marinetti was cheated by circumstance. When he preached patriotism and war, internationalism and pacificism were in vogue; and when patriotism and war were in vogue, he was upstaged by a more astute politician and a more charismatic man of letters. As it became clear, late in 1914, that public opinion was moving in favour of Italy's military intervention on the side of

* You like bowing and scraping to every foreigner, and you are disgustingly servile! You've turned yourselves into hotel waiters, guides, pimps, antique-dealers, swindlers, fakers of old masters, painters who plagiarize and copy.

Britain and France, Benito Mussolini renounced socialism and began his fateful courtship of Italian nationalism; and in the spring of 1915 Gabriele d'Annunzio returned from Paris, his refuge from menacing creditors, in order to campaign for intervention and promote himself as demagogue and national hero. There could not have been a less plausible candidate for the role. D'Annunzio was a manicured, middle-aged manikin with narrow shoulders and wide hips. Yet such was the power of his oratory that he was able to persuade huge crowds to acclaim him as Italy's man of destiny. Mussolini learnt a lot from his performances, which were quite different from Marinetti's. D'Annunzio's self-presentation was vatic rather than iconoclastic, and his orgiastic, Dionysian invocation of life owed more to Nietzsche than to Bergson. In response to his summons multitudes pledged themselves to resist barbarism and rally to the sacred cause of culture, and erstwhile Futurists prepared to defend what they had hitherto aspired to destroy. Marinetti claimed that d'Annunzio was merely following where he had led, and he tried to outposture him as great warrior, great orator, and great lover. But d'Annunzio's international prestige, his conspicuous war wound, and the *coup de théâtre* of his seizure of Fiume for the fatherland in 1919 ensured that he, and not Marinetti, captivated the post-war young. Marinetti's role in Fascist Italy was that of minor acolyte of an ideology that d'Annunzio had helped to generate.

D'Annunzio's influence softened the chauvinism of post-war Italy. His conception of national regeneration did not include the obliteration of the past, and he had no quarrel with the foreign votaries of Italy's cultural inheritance. The Anglo-Venetians, judging by Horatio Brown's comments, did not think much of him. 'What a mountebank!' wrote Brown to Rosebery. 'What a charlatan!'[37] Yet the harshest thing that d'Annunzio had to say about the Anglo-Venetians was that they were an irrelevance. 'La loro presenza', he wrote in *Il Fuoco*, 'non pesa più delle alghe vagabonde che fluttuano presso le scale dei palazzi marmorei.'* His attitude towards Venice was ambivalent. He hailed the Biennale as the prelude to a Venetian renaissance, and in *Il Fuoco* (1900) his fictional self, Stelio Effrena, enthuses about great modern warships along the Riva and hears the throb of anticipation in the Venetian silence. 'Era ovunque diffuso uno spirito di vita, fatto d'aspettazione appassionata e di contenuto ardore.'† He celebrated the foundation of Venice in an epic verse-drama, *La Nave* ('The Ship', 1907);

* Their presence counts for no more than the wandering seaweed that floats around the steps of the marble palaces.

† There was everywhere diffused a spirit of life, composed of passionate expectation and suppressed ardour.

and it was from Venice during the First World War that he addressed his rousing appeals to the nation, proclaimed the Adriatic the moral and historical inheritance of the lion of St Mark, and made his daring solo flight to bombard Vienna with propaganda. But the Dannunzian call to life and renewal was made in accents of artifice and exhaustion. In its archaic language, its elaborate stagecraft, its heavy symbolism, and its static, operatic characters, nothing is more Byzantine than that anti-Byzantine parable, *La Nave*. D'Annunzio shared with his friends Giacomo Boni and the historian Pompeo Molmenti a deep attachment to the crumbling fabric and time-worn tints of the city he always called 'Venezia la bella'. He was deeply distressed by the fall of the Campanile, and added his powerful voice to those demanding the building of a replica. Molmenti was always confident of his support in the fight against development that was vandalistic and restoration that disguised the injury and suffering of age. Both in 1919, when he resisted the proposal for a bridge for road traffic across the lagoon, and again in 1922, when he opposed the regilding and repainting with lapis lazuli of the Ca' d'Oro, he had d'Annunzio on his side.[38] In 1897, at the age of 34, d'Annunzio had abjured his Aesthetic youth; but the rest of his life was in many ways an unsuccessful attempt to live up to that renunciation. He was never an original. He made no discoveries and he broke no moulds. He was a brilliant imitator—at times a plagiarist—who wrote the scenario of his own life in language that he had borrowed from the international literary repertoire; and his youthful intoxication with the language of Decadence left a permanent addiction. Even when writing of something else, he seemed always to be writing from that enclosed, overcultivated garden where the Decadents had nurtured their exquisite pain. The work of few writers is more persistently suggestive of vitality consumed in narcissistic self-contemplation; of energy burnt out in a hard, gem-like flame. D'Annunzio's homage to Venice remained essentially the homage of a Decadent to the city where sensation countermanded action, and where passion was made morbid by remorse and nostalgia ('il rimpianto dei giorni irremediabilmente perduti'*). In *Il Fuoco* it reveals to Stelio Effrena 'la possibilità di un dolore transmutato nella più efficace energia stimolatrice';† yet at the same time it enraptures his creator with its intimations of an encumbering past; a past that accumulates in every life and drags 'like an immense dark net full of dead things' behind the ecstasy of every love affair.[39] *Il Fuoco* was planned as the first instalment of a trilogy that was to

* regret for the days that are forever lost.

† the possibility of pain transmuted into the most powerful stimulating energy.

have a heroic, Nietzschean theme of salvation through the will to endure, of triumph through creative suffering. But the two remaining novels, which d'Annunzio intended to call *La Vittoria* and *Il Trionfo della vita*, he found he could never finish.

# · VIII ·

# *Apotheosis*

THE faded frescos of Giorgione and Titian on the façade of the Fondaco dei Tedeschi were part of the Venice of decaying exteriors and veiling atmospheres that appealed to the Decadent imagination. This Venice belonged to the North and the characteristic Northern idiom of graves and worms and epitaphs. But there was a Mediterranean Venice too; a city of sumptuous interiors, where the canvases and ceilings of Titian, Veronese, and Tintoretto existed in unabated splendour, and dimness was an anticipation not of light extinguished but of radiant noon. This is the Venice that Stelio Effrena, hero of *Il Fuoco*, reveals to the *crème de la crème* of Italian society in his oration in the Ducal Palace. The scales fall from the eyes of his distinguished audience as Effrena—who is d'Annunzio's portrait of himself—expatiates apocalyptically on the Venetian painters of the climactic era, from the youth of Giorgione to the old age of Titian. 'Nessuno al mondo', he declaims,

> conobbe e assaporò meglio di loro il vino della vita. Essi ne traggono una lucida ebrietà che moltiplica il lor potere e comunica alla loro eloquenza una energia fecondatrice. E nelle loro creature più belle il battito violento dei loro polsi sembra persistere a traverso i secoli come il ritmo stesso dell'arte veneziana.*

Here d'Annunzio's mimickry registers a movement of opinion among leading European critics and art-historians. Ruskin had celebrated Venetian painting as well as Venetian architecture. He revered the Ducal Palace as much for what it contained as for how it looked. Its rooms, he wrote,

> were full of pictures by Veronese and Tintoret that made their walls as precious as so many kingdoms; so precious indeed, and so full of majesty, that sometimes when walking at evening the Lido, whence the great chain of the Alps, crested with silver

* Nobody in the world knew better than they, nor savoured more fully, the wine of life. They draw from it a lucid intoxication that multiplies their power and imparts a fertilizing energy to their eloquence. And in their most beautiful figures the violent beating of their pulses seems to persist across the centuries like the very rhythm of Venetian art.

clouds, might be seen rising above the front of the Ducal Palace, I used to feel as much in awe in gazing on the building as on the hills.

The city discovered by Walter Pater in these Venetian interiors was a city composed of 'ideal instants . . . from . . . feverish, tumultuously coloured life . . . exquisite pauses in time, in which . . . we seem to be spectators of all the fulness of existence, and which are like some consummate extract or quintessence of life'. And Bernhard Berenson, contemplating Venetian art, was conscious of 'truly Dionysiac, Bacchanalian triumphs—the triumphs of life over the ghosts that love the gloom and hate the sun'. Proust argued that this meridional, vernal Venice must be the authentic one, since it was by its art that the city was recognized: 'Et puisque à Venise ce sont des œuvres d'art, les choses magnifiques, qui sont chargées de nous donner les impressions familières de la ville, c'est esquiver le caractère de cette ville . . . de n'en représenter au contraire que les aspects misérables, là où ce qui fait sa splendeur s'éfface.'*[1]

The beauty of Venetian painting was, like the beauty of Venetian ruins, one of the cardinal rediscoveries of the later nineteenth-century. Easier travel to Venice made better known artists whose reputations had dipped following the cool judgements of influential critics like Joshua Reynolds and William Roscoe in England, Alexis-François Rio in France, and Franz Kugler in Germany. These spokesmen of evangelical generations had found much more of what they were looking for—spirituality, 'high ideas', religious feeling—in Michaelangelo and the Florentines than in the more frankly sensuous Venetians. Because the Venetian figure-painters of the Cinquecento were poorly represented outside Venice (except in Madrid, which no one visited), and because their work was travestied by engraved reproduction, the means to reappraisal were lacking until Venice itself became more accessible. Seventeenth- and eighteenth-century Venetian art remained in general disrepute; but when the Victorian student confronted Titian, Tintoretto, and Veronese on their native territory he experienced a revelation. 'If we ask ourselves', noted the 22-year-old John Addington Symonds in 1862, 'why some of the greatest Venetian masters are so little known, the answer is simple—the essence of their works cannot be represented by engraving.' Furthermore he found the Tintorettos in the Accademia in Venice a world removed from the dark, confused, and metallic canvases in England and Germany that were attributed to the master. The Venetians, he decided, were 'the giants of the world of art', and fore-

* And since in Venice it is given to works of art and magnificent things to form our prevailing impression of the city, you evade its character if you depict, contrariwise, only its miserable aspects, in which everything that makes its splendour is effaced.

most among them was Titian, 'the Sophocles of painting'. These opinions were part of a highly charged debate that had been going on ever since Ruskin's discovery of Tintoretto in the Scuola di San Rocco and his presentation of him, in the second volume of *Modern Painters* (1846), as a colossal genius and mighty soul. Henry Layard, editing the Italian sections of Kugler's *Handbuch der Geschichte der Malerei* for late-Victorian readers, defended Tintoretto against the author's strictures. 'They do scant justice', he wrote, 'to that great master, whose works are now better known and more fully understood and appreciated in England, principally through the eloquent writings of Mr Ruskin. It may be asserted with confidence that no painter has excelled him in nobility and grandeur of conception, and few in poetic intention . . .' Connoisseurs drew up a cosmic order of merit, in which the Venetian painters were given superlative ranking. Ruskin decided that Tintoretto's 'Coronation of the Virgin' (then known as the 'Paradiso') was 'by far the most precious work of art of any kind whatever now existing in the world'. Symonds divided his first prize between Titian (for the 'Assumption') and Raphael (for the 'Madonna di San Sisto'). The general popularity of Venetian art was attested by Berenson in 1894. 'Among the Italian schools of painting', he wrote, 'the Venetian has, for the majority of art-loving people, the strongest and most enduring attraction.'[2]

When you looked at Venetian architecture you became aware of the pale ghost of a great city, an afflicting emblem of time and mortality. When you looked at Venetian painting you became aware of a golden city outside time and beyond mortality, where art was paramount and culture triumphant. And you longed to enter that city, because you viewed it from a region where barbarism was setting up its kingdom.

In the later years of the nineteenth century Western intellectuals compounded their spiritual malady with political anxiety. Culture seemed threatened by liberal ideology, and much of what they wrote reads like a chronicle of disillusion with the world that liberalism had created—the world of industry, secularism, and parliamentary democracy. First in England and France, and then in Italy, Austria, and Germany, mounting evidence of moral and material crisis provoked bitter refutations of the prevailing political philosophy. Liberalism was arraigned for having replaced the fertile cult of feeling with the sterile cult of reason; for having destroyed the organic, national community and set up the corrupt, politicized state; and for having debased the currency of civilization by imposing the market economy. 'Art cannot have a real life and growth', warned William Morris, 'under the present system of commercialism and profit-mongering.'[3] In France, the Symbolists registered their revulsion from the hideous environment of the liberal bourgeoisie. The critics of democracy—Mill, Ernest Renan, Nietzsche—all feared that the rule of the majority must extinguish the influence of the noble and the wise. Bourget diagnosed a 'conflit entre la démocratie et la haute culture' as an underlying cause of the 'maladie intellectuelle' of modern France.[4] D'Annunzio voiced the disillusionment of many Italian intellectuals when, in *Il Piacere* (1888), he deplored 'the grey modern flood of democracy', which was 'submerging so many lovely and rare things' ('il grigio diluvio democratico odierno, che molte belle cose e rare sommerge miseramente'), and when, in *Le Vergini delle rocce* (1895), he wrote of the mob profaning the altars of Thought and Beauty. 'It is the same in Italy as in France', wrote Jacob Burckhardt in 1881, 'the growth of business and material things, and a marked decrease in . . . political security.' He predicted: 'One thing after another will have to be sacrificed—positions, possessions, religion, civilised manners, pure scholarship.'[5] Culture figured more and more as a frail survivor, out of its natural element and menaced either by the philistine rich or the brutalized poor. In the view of d'Annunzio and other intellectuals at odds with egalitarian, liberal Italy, it was menaced by both. Their lamentation echoed that of Matthew Arnold, who had written of Victorian England as a place where culture was crushed between the upper and nether millstones of 'materialized' wealth and 'raw and uncultivated' indigence. It was Arnold who popularized in English the term *Philister*, coined by Heinrich Heine to denote hostility to ideas, impermeability to thought.

Seen from the shadows of the modern world, Venice acquired a new and enviable prestige. As a tabernacle of civilization it was more exemplary even than ancient Greece, whose credentials were being blemished by anthropological research. The plight of art was something it had never known, because neither the brutalized poor nor the philistine rich were a feature of its history. According to Ruskin, Venice had possessed the ideal proletariat. In 'The Nature of Gothic', the famous central chapter of the second volume of *The Stones of Venice*, he depicted a utopia where labour had been made happy by thought, just as thought had been made happy by labour. The Venetian workman had not been a mental slave, an envious drudge, but a free artist motivated by the joy of creation. And modern visitors were reminded of this ideal proletariat, to whom art belonged and who belonged to art, when they recognized in the common people figures from the canvases of the great Venetian painters. Northerners were in the habit of seeing art in the popular life of the South. Such recognition was one of the joys of Mediterranean travel. But common life was nowhere so obviously associated with art as it was in Venice. Anne Thackeray found the Rialto crowded with figures from celebrated pictures: 'Virgins went by, carrying their infants; St. Peter is bargaining his silver fish; Judas is making low bows to a fat old monk . . . Titian's mother, out of the "Presentation", who was sitting by with her basket of eggs, smiled and patted the young Madonna on her shoulder.' Julia Cartwright described how a carpenter's shop became the realization of an old master. Everything was 'just as in that famous Annunciation which you saw the day before in San Rocco yonder . . . It is all so like Tintoret's picture that you almost expect to hear the rustle of angel-wings in the air above.' The face of the working-class heroine of Ouida's story *Santa Barbara* is a replica of that painted by Palma Vecchio, 'whilst her throat and bosom and arms [are] those of Veronese's Europa'. Among the gondoliers Elizabeth Eastlake and Horatio Brown recognized 'the eager figure of the classic charioteer urging his coursers in the antique race' and 'Bacchus stepped from Tintoret's loveliest picture'. Fishermen too were denizens of the world of art. The Revd Stopford Brooke, an Edwardian visitor who liked to observe 'the Venetian youths of the people', discovered with a thrill that

> to look at one of these young Venetian fishers, standing in the blaze of the sun, with the greenish water glistening around him, its reflections playing on his glowing limbs, and all his body flaming soft as from an inward fire, is to see the very thing which Giorgione painted on the walls of palaces . . . which Titian and Tintoret laid on their canvas and emblazoned on their fresco. . . . There is a young and naked St. Sebastian by Titian in the Salute which might stand for one of the fishers of the lagoon.

In Proust's novel it is a young pedlar of glassware who merits the old-master attribution. 'La beauté de ses dix-sept ans était si noble, si radieuse, que c'était un vrai Titien à acquérir avant de s'en aller.'*[6]

Others discerned the ideal plutocracy—a patriciate of connoisseurs in symbiotic relationship with creative genius. 'Il y a dans la vie de chaque artiste, peintre, sculpteur ou architecte', wrote Charles Yriarte, 'un nom de patricien qui rayonne comme celui d'un patron bien faisant, un protecteur généreux et bienveillant . . . auquel l'artiste inspiré rend plus tard, en gloire et en immortalité, le généreux appui qu'il sut donner à sa jeunesse, à son obscurité, et à son génie naissant.'† Symonds wrote of 'a prudent aristocracy who spent vast wealth on public shows and on the maintenance of a more

* The beauty of her seventeen years was so noble and so radiant that she was truly a Titian to be acquired before leaving.

† There is in the life of every artist—painter, sculptor, or architect—the name of a patrician, which shines like that of a kind patron, of a generous and benevolent protector, to whom the inspired artist later pays back, in the form of glory and immortality, the generous support given to youth, obscurity, and burgeoning genius.

than imperial civic majesty'; who employed great painters 'to invest the worldly grandeur of human life at one of its most gorgeous epochs with the dignity of the highest art'. Art and commerce had enhanced each other in perfect connubial harmony. The lesson to be learnt from Venice, said Molmenti, was that 'il sentimento dell'arte non uccide l'industria; il commercio non è d'impaccio all'operosità. . . . L'idea del bello si può unire a quella dell'utile.'* It was a lesson that the *Builder* drove home in a leading article:

> Those who built Venice, as it now stands, were men who had the greatest respect for and made the most brilliant successes in trade and manufacture. . . . The archives of ancient Venice show conclusively that so far from mercantile and commercial enterprise and industry being incompatible with art, the two ends were pursued by the Venetians with equal devotion and success.[7]

So in Venice art that had few friends contemplated art that had had no enemies. Ruskin perceived 'a world from which all ignoble care and petty thoughts were banished, with all the common and poor elements of life'.[8] The vision was inspiring, intoxicating; but at the same time so different, so remote from modern conditions, that it suggested conflicting inferences and drastic remedies. Love of Venice therefore meant political passion, as well as literary self-absorption. It was a symptom of the Barrèsien 'Culte du Moi'; of d'Annunzio's autoeroticism; of a whole epidemic of *luxe, calme, et volupté*; yet it was also a stimulant to an urgent concern to set the world to rights. The yearning for withdrawal into an ideal city of the mind, for disengagement from the mundane and the provisional, translated into politics of anger and politics of fear. William Morris and Philip Webb were revolutionary socialists, working for the overthrow of property, profit, and competition. They looked to a social order that would transcend class and nation. But neither Barrès nor d'Annunzio, neither Wagner nor Nietzsche, acquired a taste for popular causes that conquered his contempt for popular taste. So these Venetophiles became psychotherapists to the fearful. They ministered with art and philosophy to that cult of redeeming hero and transcendent nation that was the other antidote to liberalism.

The angry believed that art was popular; that it was made and appreciated by the many. 'When men say popes, kings, and emperors built such and such buildings', argued Morris, 'it is a way of speaking. You look in your history-books to see who built Westminster Abbey, who built St Sophia at Constantinople, and they will tell you, Henry III, Justinian the Emperor. Did they?

* the artistic sensibility does not kill industry; commerce is no impediment to craftsmanship. . . . The idea of the beautiful can co-exist with that of the useful.

or, rather, men like you and me, handicraftsmen, who have left no names behind them, nothing but their work.' The fine arts (what the Victorians called 'High Art') had their roots in this popular art, and when popular art was ailing they became 'nothing but dull adjuncts to unmeaning pomp, or ingenious toys for a few rich and idle men'.[9] The art of the past was great because it was 'organic'—that is to say, the people had shared in the creation and the enjoyment of it. In modern times art was languishing because the people had been diverted from creation and enjoyment to a sordid struggle for survival. Only a socialist revolution therefore could ensure the rebirth of art and the salvation of the artistic heritage. 'It will not be possible', Morris told the SPAB in 1884,

> for a small knot of cultivated people to keep alive an interest in the art and records of the past amidst the present conditions of a sordid and heart-breaking struggle for existence for the many, and a languid sauntering through life for the few. But when society is so reconstituted that all citizens will have a chance made up of due leisure and reasonable work, then will all society, and not our 'society' only, resolve to protect ancient buildings . . . for then at last they will begin to understand that they are a part of their present lives, and part of themselves.[10]

The fearful believed that art was esoteric, rooted not in common life but in exceptional sensibility. It was what d'Annunzio called 'a magnificent gift of the few to the many'; what Yeats celebrated as 'the inherited glory of the rich'. Socialism would therefore be not its salvation but its ruin. The redistribution of property would cast pearls before swine. 'You none of you know as yet', Burckhardt had warned his friends in 1846, 'what the people are, and how easily they turn into a barbarian horde. You don't know what a tyranny is going to be exercised on the spirit, on the pretext that culture is the secret ally of capital and must be destroyed.'[11] Forty years later there was no lack of converts to his way of thinking. Throughout Europe the move to democracy provoked a fear that socialism must follow, and there was an intellectual reaction which desiderated the rule of an élite, and which in France, Germany, and Italy fed a strain of messianic fever into political thinking. The less hectic but no less anxious Anglo-Saxon response is illustrated by *The Princess Casamassima*, a novel of social realism that Henry James published in 1886. This is very much the work of a frequenter of country houses and Italian *palazzi*, who senses that refined ancestral comforts exist in close proximity to 'some sinister anarchic underworld, heaving in its pain, its power, and its hate.' The graces of culture, which Venice incomparably represents, are under threat from the political left, because unlike power and property, they cannot be redistributed. When culture is shared, it is destroyed. James's hero, the artisan Hyacinth Robinson, realizes that Venice is at risk from the activities of the arch-revolutionary Hoffendahl, since 'he would cut up the ceilings of Veronese into strips, so that everyone might have a little piece'. Aristocratic privilege, and privation among the many, were the price that had to be paid for the exquisite things of life. Without the suffering of the disinherited, there would be no inheritance. At the centre of the novel there is a choice between revolution and culture, between social justice and art. And Hyacinth's dilemma is doubly painful, because he is himself a divided hero. The posthumous son of an English nobleman by a French *fille de joie*, he personifies the conflict between aristocratic refinement and proletarian uncouthness. 'There was no peace for him', we are told, 'between the two currents that flowed in his nature, the blood of the passionate plebian mother, and that of his long-descended super-civilised sire.' In London he associates with anarchists and revolutionaries, and pledges his life to 'the great rectification'; but abroad he is deconverted by the revelation of art. What becomes supreme in his mind in Paris is 'not the idea of how the society that surrounded him should be destroyed', but rather 'the sense of the wonderful things it had produced, of the fabric of beauty it had raised'. In Venice he realizes that he is capable of fighting for 'the splendid accumulations of the happy few . . . the monuments and treasures of art, the great palaces and properties, the conquests of learning and taste, the general fabric of civilisation as we know it'. So rather than redeem his pledge and carry out the assassination that will subvert the social order, he becomes a martyr to culture by turning his pistol on himself.

Neither the critics nor the reading public liked *The Princess Casamassima*—partly because it was too plausible to be entertaining. Its assumptions about the hereditary nature of taste and sensibility were corroborated by the new science of eugenics, and there was an unwelcome truth to life in its depiction of the political left as antagonists of art. Morris and Webb and the late-Victorian and Edwardian socialists were more concerned about the 'decorative arts'—that is, handicrafts and the ornamental element in architecture—than about the fine arts. There is little evidence of 'High Art' in the Utopia of Morris's *News from Nowhere*, and both Morris and Webb were very selective in their admiration of Venice. Neither had any feeling for what Webb called 'the cold and formal architecture' of the Venetian Renaissance. The corollary of their perverse belief that majorities make culture, was the idea that minorities make civilization. Coleridge had defined 'civilization' as 'the hectic of disease, not the bloom of health'; and Morris was clearly thinking in Coleridgean terms when he wrote, in 1885, that civilization was doomed to destruction. 'What a joy it is to think of it!'

he added. He presumably wanted and expected Renaissance Venice to perish in the holocaust, because nothing more obviously belonged to 'civilization' than the legacy of the Venetian oligarchy. Such, in essence, had been Ruskin's argument against it; and Ruskin's strictures resurfaced in the reactions of George Bernard Shaw. 'Somehow', wrote Shaw to Morris in 1891, 'there is a painful element in the whole affair which throws me back on my old iconoclastic idea of destroying the entire show. For it is a show and nothing else.' In an outburst of intolerance, he went on to repudiate even the principles of the SPAB:

> I do not believe that a permanent, living art can ever come out of the conditions of Venetian splendour, even at its greatest time. The best art of all will come when we are rid of splendour and everything in the glorious line. Then by all means let them restore and ruin and cut up missal papers and scrape and do what they like. The old bricks are of no use in building up the new art, as far as I can see.

In the world of French politics likewise, red seemed to mean danger for the treasures of the past. Although some French socialists—Jean-Léon Jaurès, most notably—argued that in a socialist society art would be in its natural element, others were less reassuring. The anarchist Jules Félix Grandjouan went so far as to denounce the Louvre as the temple of an oppressing class, and in a rabid article of 1908 he prophesied its destruction by the outraged poor.[12]

This sort of language encouraged the belief that precious things were at risk not only in practice but in theory. It appeared that the obliteration of art was not a price that the just society had to pay, but a condition that it actually required. Culture was confronting not barbarian instinct, but barbarian instinct sanctioned by perverse intelligence—what Turgenev called 'nihilism'. *The Princess Casamassima* incorporated these anxieties; and if the novel suffered at first because they seemed only too well-founded, it suffered subsequently because they seemed half mistaken. The ideas that most vivified the novel in its day were precisely those that diminished its power of survival. Eugenics went out of date, leaving it stranded among obsolete notions about the heritability of 'aristocratic' and 'plebian' mentalities. And then history took an unexpected turn, emptying its low-life atmosphere of any real sense of menace and making its tragedy look like melodrama.

The case of Venice showed quite clearly that the danger to art was not explicable in terms of a simple polarity between privilege and privation. The threat to Venice did not come from 'the people'; nor was concern for the city limited to an élite of birth and intellect. The threat to Venice—the human threat that is—came mainly from a cadre of architects, engineers, town-planners, and sanitary scientists who were motivated by professional ambition rather than by political ideology. The campaign for conservation on the other hand was assisted by a wide range of public opinion. It transpired that Demos did not favour the destruction or even the transformation of Venice. Demos, on the whole, preferred it as it was.

The new perception of Venice, its adoption by the cultivated as a unique and priceless inheritance, coincided with the commercialization of leisure. In the last decades of the nineteenth century the tourist industry needed shrines for its secular pilgrims to visit and relics for them to venerate. The publishing and entertainment industries needed glamorous places to bring to audiences who were yearning but unable to travel. And the richness of Venice in shrines, relics, and glamour was attested by all those tributes of literature and art that had been accumulating since the original secular pilgrim, Byron's Childe Harold, had stood on the Bridge of Sighs and experienced the numinous. Before the railway age Byron, reprinted in Murray's handbooks, had commanded the almost undivided attention of tourists. His poetry had told them how to behave and what to admire.[13] By the middle of the century he was in eclipse. Matthew Arnold claimed that he was no longer read, and Chateaubriand discovered that the plaque commemorating his residence had disappeared from the façade of the Palazzo Mocenigo.[14] But then he was revived, as part of the cultural packaging that was contrived to sell Venice to the huge new travelling public. Byron was recruited, together with Ruskin, to inspire and instruct the railway tourists. The Venetian works of these two were served up whole or, more often, in extracts, together with other bits and pieces from the Venetophile canon. This was the age of the traveller's anthology—books like Augustus Hare's *Venice* (1884) and Alfred Byatt's *The Charm of Venice* (1908). The *cognoscenti* half despised the customers of Thomas Cook and Arnold Lunn, but at the same time recognized the importance of their hearts and minds in a world swayed less by the quality than the quantity of opinion. So when the first organized parties appeared in Venice they were pampered and tutored by the notables. In 1889 a group of sixty artisans, clerks, salesmen, and schoolteachers from Toynbee Hall, the social settlement in east London, was entertained to tea at the Rezzonico by Pen Browning and lectured by Symonds and Horatio Brown. Parties from the Artworkers' Guild in 1891, and again from Toynbee Hall in the following year, were given similar treatment.[15] Among them was an artisan who had come straight from that London of anarchists and slums that James had described in *The Princess Casamassima*. Unlike Hyacinth Robinson, he had no aristocratic blood to refine him; nevertheless his response to the treasures of civilization was every bit as exemplary. This was Thomas Okey, a basket-weaver from Spitalfields. Okey, educated in his spare time by French and German refugees and by University Extension lecturers at Toynbee Hall and the Working Men's College, wrote two books on Venice, translated Dante and St Francis, and became professor of Italian at Cambridge. His achievement was exceptional; but his reverence was not. The world was accepting Venice at the Venetophiles' evaluation. Their ideas were shaping popular perception; their language was being mimicked in media platitudes and commercial clichés. An esoteric conceit, a vagary of Romanticism in its feverish decline, was acquiring the permanence and potency of myth.

Those too poor to go to Venice now had Venice brought to them. Good books had never been so cheap. Leonard Baste, reading Ruskin at the beginning of *Howards End*, has obviously bought the Everyman edition of *The Stones of Venice*, which was published in 1907. Venice reconstructed on the stage dazzled theatrical London. For the price of a seat in the gallery you could buy the illusion of travel. There were two revivals of *The Merchant of Venice* in the 1870s, both with elaborate and authentic décor. Sir Squire and Marie Bancroft, lessees of the old Prince of Wales's Theatre in Tottenham Street, visited Venice in 1874 with their chief scenery-painter George Gordon to select locations for their production of Shakespeare's play. Gordon then reproduced these in act-drops and stage-sets which marked the beginning of an era of 'archaeological' realism in theatre design. Lady Bancroft recalled that for the first act, located under the arcades of the Ducal Palace,

> elaborate capitals of enormous weight, absolute reproductions of those which crown the pillars of the colonnade of the Doge's Palace, were cast in plaster, and part of the wall of the theatre had to be cut away to find room for them to be moved, by means of trucks, off and on the small stage, which, though narrow, fortunately had a depth of 38 feet.

The costumes were copied from the pictures of Titian and Veronese, and Ellen Terry, who played Portia, reckoned that 'a more gorgeous and complete little spectacle had never been seen on the English stage'. The critics were not impressed by the Bancrofts' approach to Shakespeare. Joseph Knight, writing in the *Athenaeum*, disapproved of attempts to 'convert his plays into spectacular entertainments'. However, he agreed that these would suit 'ignorant pleasure-seekers', and had to admit that as an evocation of Venice the Bancrofts' production was remarkable: 'Superb views of Venice are presented. The gay, idle *insouciante* and withal mysterious life of the

Queen of the Adriatic is depicted with as much truth and colour as in the pages of *Consuelo*. . . . The busy masque of medieval Venice defiles with marvellous fidelity before our eyes.' The sets for Henry Irving's production of *The Merchant* at the Lyceum Theatre in 1879 were based on prints and photographs that Irving had collected in Venice. Knight judged them 'at once striking, natural, and unobtrusive'. Ellen Terry again played Portia, 'got up in exact imitation of those stately Venetian dames who still gaze down from the pictures of Paolo Veronese'. In 1908 a new generation of theatre-goers saw what was probably the most lavish staging of the play in theatrical history. It was produced at His Majesty's Theatre by Beerbohm Tree, who presented Shakespeare with huge crowds, realistic sets, and technical effects.[16] Meanwhile in 1892 the impresario Imré Kiralfy, who specialized in epic events that combined entertainment with education, had brought his own version of Venice to the people of London. At Olympia, the huge exhibition hall in Hammersmith Road, he built a replica of part of the city in fibrous plaster, complete with palaces, cafés, canals, gondolas, gondoliers, and what *The Times* described as 'an intricate labyrinth of bridges, footways, and landing stages'. Twice daily, at 2.30 and 6.30 every day for most of the year, all this became the setting for a spectacular show called 'Venice the Bride of the Sea'. Public interest was immense. In June there were over 30,000 visitors on one day alone, and the first edition of Horatio Brown's *Life on the Lagoons* very quickly sold out. 'Venice at Olympia' marked the adoption of Venice as a totem by the mass-entertainment industry, and added enormously to the pressures that were forcing it to conform to well-established preconceptions and expectations.[17]

Some Venetians resented this consecration of their city; this pre-emption by international opinion of the right to define its role and control its destiny. Like the Futurists, they disparaged tourism as economically uncertain and morally vitiating, and there was some sympathy for their feelings abroad. In 1887 the *Builder* warned the British public that they had 'no right to require [of] the inhabitants of any old city that they should be content to reduce themselves to the condition of the custodians of a museum'. The authoress Margaret Oliphant, among others, drew attention to the fact that resistance to change was alien to every Venetian tradition. 'We cannot doubt', she wrote, 'that the Michiels, the Dandolos, the Foscari, the great rulers who formed Venice, had steamboats existed in their day . . . would have adopted them without question.' But such arguments underestimated the importance of tourism in the modern Venetian economy. Tourism was creating a large lobby of Venetians whose interests were served not by the historical city of the doges but by the mythological city of literature and art. In order to prosper they needed to satisfy tourists; and in order to satisfy tourists, as a correspondent of *The Times* wrote in 1887, 'they need[ed] but let the place alone, for the ideal wish of thousands [was] to glide down the Grand Canal, past palaces and Rialto, or to cross the lagoon to [the] Lido, with Byron and Shelley'.[18]

Hindered by these countervailing pressures, the progressivists' achievement was modest. Traditional industries were encouraged and a few new ones introduced. The port was modernized and enlarged, and the urban landscape modified. Several new roads were built; some canals were filled in; and the south-western tip of the city was transformed by the extension of the railway across the Campo di Marte and by the construction of a *cotonoficio* (cotton factory) on the remaining section of the old *àrzere*, or embankment, of Santa Marta. Venice altered during the thirty years before the First World War—but not beyond recognition. The rate of change was probably slower than it had been under the Austrians, and was certainly slower than it had been under the French. Venice underwent nothing comparable to the *sventramento* (evisceration) of Rome and Milan. Visitors who returned after a long absence had little difficulty in recognizing the place they had known. When the painter Thomas Rooke, who had been a member of the artistic colony in Venice in the 1870s, went back in 1907 he experienced no sense of dislocation. 'Venice, to our great joy,' he wrote to Sydney Cockerell,

> proves to be in no sense the baseless dream of our youth. Though there are many things changed and factory chimneys even begin to take their place among the towers, we feel we are still in the only ancient city we have ever been in, and that perhaps for a hundred years yet it will still be that to those who can get here.[19]

Of all the major cities of Italy, this one probably changed least. When in 1886 the *sindaco* Serego degli Alighieri presented to the Consiglio Communale a comprehensive scheme of slum-clearance and redevelopment (the *Piano Regolatore di Risanamento*), Molmenti, supported by Boni, organized a concerted and effective opposition. 'Hanno le loro esigenze', cried Molmenti, 'l'igiene e la decenza, ma quando si afferma che Venezia non potrà risorgere fino a che le vetture non correranno per alcune strade, si pronuncia una balorda bestemmia.'* Molmenti's philippic, 'Delendae Venetiae' which was published in the *Nuova Antologia*, alerted the authorities in Rome, and an official commission of inquiry was set up under Camillo Boito. This drastically curtailed Alighieri's scheme. In the

* Hygiene and decency have their claims, but it is a stupid blasphemy to assert that Venice will never revive until carriages are running through some of its streets.

event only sixteen out of his forty-one proposals were implemented. From now on the Consiglio Communale paid as much attention to conservation as to modernization. In 1890 it bought the *squero* at San Trovaso, the oldest and most famous of the boatyards where gondolas were made and repaired, together with the surrounding houses and gardens, in order, as Boni told Webb, 'to prevent that a spot so characteristic of old Venice should fall into the hands of speculators'. A second bridge across the lagoon was mooted, but Molmenti led a successful campaign against it. Venice remained inaccessible to traffic until 1933, when a road that had been attached to the railway viaduct was opened. Even then the automobile got no further than the periphery. All plans designed to give vehicles access to the centre of the city—including that of filling in the Grand Canal—came to nothing, and probably never stood a serious chance of being adopted. In 1889 the *Builder* had suggested that there was an obvious way of reconciling modernity with preservation. 'Surely' it argued, 'it would be better to build a "New Venice" on another site, and leave us the old one as a thing of beauty.' In the twentieth century this idea was in effect adopted. Industrial and residential development was diverted to the Lido and, after 1917, to Porto Marghera in the area of Mestre on the mainland. The foundry and factory on Sant'Elena were replaced with working-class housing, and the old city, now designated 'il centro storico' ('the historic centre') was left in comparative peace—a counterfeit but venerated relic, enclosed on one horizon by a garish suburban seaside, and on the other by a grey industrial megalopolis.[20]

So as a political parable of aristocratic Beauty and proletarian Beast *The Princess Casamassima* was soon made redundant by the complexity of events. But to see the novel thus is to see it with one eye closed. In its stereoscopic depth it has other dimensions. James was a leading practitioner of the 'psychological' novel; of 'the analytic fashion of telling a story', which painted characters from the inside and uncovered 'motives, reasons, relations, explanations'. 'There are few things more exciting to me', he wrote, 'than a psychological reason.'[21] In *The Princess Casamassima* he reveals a Venice at risk not only from a social and political underworld, but from a psychological underworld too. He exposes the malefic forces beneath the surface of polite society; and as a study of a civilization threatened with destruction by the human perversity of its aristocratic rulers, the novel was more than merely plausible.

James depicts an aristocracy that is not aristocratic. Hyacinth confronts in the memory of his own father 'a nobleman altogether wanting in nobility'; and through his association with two aristocratic women he comes to understand that park gates, separating the world of 'freedom and ease, knowledge and power, money, opportunity and satiety' from the mob, are no barriers against destructive instincts and vulgar emotions. Both the Princess Casamassima and Lady Aurora Langrish are revolutionaries. Both associate with the proletariat, forswear luxury and fine living, and work for a cause that will overturn the established order and inaugurate social justice. Both therefore betray the cause of civilization and of art. But they are doubly culpable, because their credentials are false. They are not what they claim to be. They are motivated not by compassion or by social conscience, but by private grudges and sexual frustration. They are excursionists in the world of the deprived; picnickers among the slums. They retain, and use, the option of retreat into the world of fashion and comfort. The princess is one of James's pampered and power-hungry women: the ruthless casualty of a broken marriage who uses the misfortunes of the poor as a weapon in her vendetta against the husband she despises. Her revolt illustrates the psychology of wounded vanity and thwarted egotism. The revolt of Lady Aurora illustrates the psychology of frustrated passion. Ungainly, unendowed, and undesired by the young revolutionary whom she loves, she is a case of 'awkward aristocratic spinsterhood' whose pent-up infatuation expresses itself as 'vengeful irony'. 'I want to live!' she cries; and she finds a consolation for not living in abetting the undermining of the high society that her family adorn. These women are oblivious of the *raison d'être* of aristocracy. They are not martyrs, but makers of martyrs; and what divides them from Hyacinth, the true aristocrat, is their inability to be magnanimous.

As a novel about low passion in high places, about rank and power allied with barbarian irresponsibility, *The Princess Casamassima* grew more topical, not less. The consequences that flowed from the assassination of an Austrian archduke in June 1914 reminded Europe that its fate was still determined by chancelleries and dynasties whose pedigrees were aristocratic but whose minds and characters were tragically commonplace. And there could be no more convincing evidence than the fate of Venice in the First World War that art was in peril from a nobility 'altogether wanting in nobility'.

It was clear that if Italy entered the war on the side of Britain and France, Venice would be exceptionally vulnerable. It was within the range of naval guns, and its nearness to the Austrian frontier exposed it to the risk of bombardment from the land as well. The ruins of Ypres, Arras, and Reims testified that nothing was too precious to be a target for artillery in modern war; so precautions were being taken as the secret negotiations were being conducted that issued in the Treaty of London of April 1915. Well before it began to fight, the Italian army was evacuating art treasures from Venice. Ugo Ojetti, made a captain in the Engineers, supervised the removal of hundreds of pictures from churches, palaces, and the Accademia. Enormous canvases like Tintoretto's 'Coronation of the Virgin' and Titian's 'Presentation' were wrapped around cylinders 60–80 centimetres in diameter and 6–10 metres in length. Titian's huge 'Assumption', painted on a single panel, had to be taken to Cremona by water, since it was too big, when encased, to pass through any railway tunnel. Vivarini's stained glass was removed from SS Giovanni e Paolo. The bronze horses were lowered from the terrace of San Marco, transported by water to Cremona, and thence by rail to Rome. Fragile monuments were enveloped in sandbags and seaweed mattresses.

On 24 May 1915, the day that Italy declared war, Venice was bombarded neither from the sea nor from the land, but from the air. This was the first of forty-two Austrian raids. Their intensity and duration increased as the war went on, and by 1918 incendiary devices had been replaced by bombs packed with high explosive. The attacks were carried out by daylight or moonlight, and initially lasted about an hour; but on 9 August 1916, following the Italian capture of Gorizia, the Austrians retaliated with a raid of several hours. On the night of 26/7 February 1918 fifty aircraft attacked the city in relays for eight hours, from ten at night until six in the morning. They discharged some 300 bombs, each containing more than 300 kilos of explosive. In all about a thousand bombs were dropped on Venice. At first some attempt was made to target strategic points—the railway station, the arsenal, the cotton mill. But the later raids were indiscriminate and in any case, given the height at which the aircraft flew and the dense construction of the city, accurate aim was impossible. So even though many of the bombs fell harmlessly into the canals and the lagoon, the damage inflicted was serious and the risks incurred by the population were considerable.

For the first few months there was little evidence of panic and life went on almost as usual. At night in the Piazza the café tables were still crowded, even though tourists had disappeared and the prohibition of light was strictly enforced. Many Venetians preferred to sit in the open in the dark rather than endure the stifling heat of rooms behind thick blinds. It was generally thought that outside was safer, too. But when the raids became more frequent and more deadly, and the cold weather kept people indoors, Venice was a dangerous place for civilians. The magical nights became interludes of terror, when sleep was torn apart by wailing sirens and crowds of screaming women and children flocked to the buildings requisitioned as shelters. 'Signora', wrote Angelo Fusato, her father's old gondolier, to Madge Vaughan in January 1918,

> può immaginarsi la povera mia moglie in quali condizioni si ritrova, di piangere giorno e notte. Si può immaginare come ci troviamo qui a Venezia, che ora non è più la città che era, essendo che si vive di miseria e di paura, ché di notte dobbiamo alzarci dall'letto per motivo dei veloplani nemici che viene [sic].*[22]

Three buildings serving as hospitals were hit, and many houses in the remoter, poorer parts of the city were demolished and their inhabitants either killed or wounded. In the Campo dei Mori, in Cannaregio, a plaque marks the spot where an Austrian missile fell in August 1917, killing fourteen people. The total of human casualties was sixty dead and more than eighty seriously wounded. As military statistics the figures were trivial; but the air raids on Venice proclaimed a fact whose direness was not measured by statistics. Culture no longer had a sanctum or a citadel. Anarchy to whom nothing was sacred had acquired a weapon from which nothing was immune. This was Europe's first experience of *Blitzkrieg*. Its sequel was to be the bombing of Cologne and Dresden and the 'Baedeker raids' on English cathedral cities in the Second World War.

Material wreckage was significant. The cotton factory was ruined and eight churches and four *palazzi* more or less seriously damaged. The church of the Scalzi, near the railway station, was one of the first and gravest casualties. It was hit on the night of 25 May 1915. The pavement and marble decoration were mutilated and Tiepolo's ceiling, depicting the translation of the Casa Santa, was destroyed. When Barrès visited the church the following year he found a gaping hole where the roof had been and nothing left of Tiepolo's masterpiece save a pile of plaster-dust in a corner of the chapel. On 9 August the roof of Santa Maria Formosa was burnt out, and on the following day the lantern of S. Pietro di Castello was destroyed and the cupola injured. Less than a week later a bomb landed near the base of the campanile of S. Francesco della Vigna. It exploded in the ground, blew in the sacristy wall, and cracked the foundations. On 10 September 1916 San Marco had a narrow escape when an incendiary bomb fell only a few yards from the main portal. During the midnight raid of 11/12 September 1916 a missile which pierced the southern clerestory roof of SS Giovanni e Paolo exploded inside the church, demolishing the north clerestory wall, stripping away all the plaster, blackening and fracturing Piazzetta's ceiling

* Signora, you can imagine what a state my poor wife is in, crying all day and all night. You can imagine how things are for us here in Venice, which is no longer the city it was. We live in misery and fear, because the enemy planes come and turn us out of bed at night.

fresco of St Dominic, and blowing out all the glass. The capacity for damage was far in excess of the actual amount caused. Given the uncertainty of aim in the earlier raids, and the random nature of the later ones, there was never an assurance of safety for any building, however celebrated. 'Neither the Venetians nor their enemies', said Horatio Brown, 'can tell what precious monument may not be sacrificed in some future raid.' This use of air-power could easily have caused the loss of architecture that Europe had agreed to venerate as one of its supreme achievements. The daring and distinctive design of the Ducal Palace, with light colonnades carrying heavy masonry above, meant that it would have collapsed if any of the angle-columns had been blown in. The roof was a forest of dry timber which an incendiary bomb would have set ablaze. The roof of San Marco, likewise, was highly vulnerable. If a bomb had fallen on any of the cupolas the outer shell of lead and wood would certainly have been destroyed, and it was unlikely that the fragile inner cupola of brick, carrying priceless mosaic, would have resisted the impact.

Attempts were made to minimize the risk, and by the end of 1916 the principal buildings of the Piazza were wearing weird protective carapaces. The angles of the Ducal Palace were sheathed in brick towers some 6 metres high and a metre thick; the openings of the lower colonnade were filled with cement blocks; and the upper colonnade was fortified with heavy baulks of timber. 'It looks', commented Horatio Brown, 'as though the whole place were trying to turn itself into a medieval fortress.' The entire façade of San Marco had disappeared behind a double screen of timber, which was filled with sandbags and made fireproof with asbestos and cement. Anti-aircraft batteries, set up on the Lido and on the islands of the Lagoon, tried to deter the aggressor with artillery fire and searchlight beams. But no enemy plane was shot down, the raids did not abate, and Venice remained threatened as never before.[23]

It was a part of the post-war legend of the Great War that this had been a sudden and traumatic ending of mythologies. *Avant-guerre* denoted a golden age. The summer of 1914 became the last hour of innocence, and the conflict a seismic chasm in history's thoroughfare. It divided a present which was a wasteland from a past which was a land where it was always afternoon. That was the legend. In fact, the mythologies had perished before the war began, and science was ready with explanations for the nightmare. Henry James belonged to an intelligentsia whose belief in a divinely ordered, benevolent universe had been wrecked by Darwin and whose understanding of human behaviour was being revolutionized by anthropology and experimental psychology. Man the Creator's masterpiece had become man the accident of nature—a stranger to God and at war with himself. The demon that could be overcome, the beast that submitted to the conquering angel, had reappeared as a dark primitive self that would not be gainsaid. Religion and philosophy had become random growths in the wilderness of personality; the world of art and morality had become a world of racial fetishes and tribal customs. Total war was something that this well-developed matrix of ideas could accommodate. But in one of its aspects the war signified not the obsolescence but the vitality of mythology. If the danger to Venice had not been real, it would have been necessary to invent it, because plight and peril had become essential to its post-historical existence. In assaulting the city that had become a paragon, the Austrians were conforming to a myth that had flourished, not failed, amid the rank disarray of the nineteenth-century mind. An idea antedated the event; and the event found expression in a well-worn literary idiom:

> Du rivage du Lido, des îles, de tous les points de la Lagune, le tir de barrage éclate contre l'invisible ennemi qui s'approche et qui rôde, ailé et sinistre, autour de la magnifique proie qu'il convoite de dominer de son vol néfaste. . . . Que de fois ma pensée inquiète n'est-elle pas allée vers Venise menacée, en attendant l'heure où elle serait enfin Venise sauvée! Bien souvent je me suis fait conter la vie de la belle assiégée. . . . La beauté même de Venise était une cible.*[24]

By describing the raids in these terms, Henri de Régnier was identifying them as the fulfilment of a destiny that had already been willed by the collective imagination. As archetypal victim, post-historical Venice had become one of the most potent symbols in a prevailing aesthetic of pain, endurance, and terrible beauty. In the Decadent recension it figured as the passive prey of plague, time, flood, and human perversity. To Nietzsche it expressed the dynamism of striving and perseverance: the will to exist and to triumph that is generated by adversity.[25] So by dropping their bombs on Venice the Austrian High Command were not merely improvising a military response. Nor were they simply yielding to primitive impulses of anger, fear, and hatred. No doubt they were doing all these things; but they were at the same time acting a role that had already been written into a modern scenario of crucifixion and redemption. Venice featured in that scenario as the city transfigured by suffering and imminent catastrophe. It became superlative as its fate became afflicting, and by subjecting it to the martyr-

* From the shore of the Lido, from the islands, from all quarters of the lagoon, the barrage-fire erupts against the invisible enemy who approaches and prowls, winged and sinister, around the magnificent prey whom he lusts to dominate in his evil flight. . . . How many times did my anxious thoughts not go out to Venice in peril, in anticipation of the time when it would at last become Venice preserved! Often did I make them tell me about the life of the fair besieged. . . . The very beauty of Venice was a target.

dom of modern war the Austrians completed its apotheosis by perfecting its beauty. Observers who saw it under wartime conditions experienced a rich and rare epiphany. 'Whoever has not seen Venice now by moonlight', wrote Horatio Brown, 'with no artificial light to challenge and confuse the purity, strength, and efficacy of that burnished-silver screen upon Istrian stone . . . has not seen Venice at all.'[26] G. Ward Price, correspondent of *The Times*, agreed. 'Never before', he reported,

> has her beauty been of such rare purity as in these fateful days when from the Rialto Bridge you can hear the enemy guns growling for the city as their prey. A dreamy silence rests like an enchantment on the place, for the gravity of the hour has banished all the modern incongruities that peacetime exploitation had grafted there, and Venice seems to be brooding wistfully over the long and splendid pageant of memories that make her past. . . . The black gondola glides through a city more beautiful in the silence and stillness of this war-time trance of hers than ever in the fulness of her vivacious life.[27]

As science debunked scripture, deified matter, and claimed an ever wider territory as its domain, art too became more strident, more imperious in its pretensions. It gave up serving God, and aspired instead to usurp the throne He had been forced to abdicate. One after another writers challenged the idea that history had created mind. Wilde, Henry James, Clive Bell, Yeats—all endorsed the view that mind had created history, by acts of the imagination. Their pronouncements resounded like religious dogma: 'Life imitates art far more than art imitates life.' 'It is art that *makes* life, makes interest, makes importance.' 'Though art owed nothing to life, life might well owe something to art.' 'Death and life were not/Till man made up the whole/Made lock, stock, and barrel/Out of his bitter soul.' Guillaume Apollinaire sketched a new cosmic dispensation, in which the artist was demiurge:

> Sans les poètes, sans les artistes . . . tout se déferait dans le chaos. Plus de saisons, plus de civilisation, plus de pensée, plus d'humanité, plus de vie même, et l'impuissante obscurité régnerait à jamais.
>
> Les poètes et les artistes déterminent de concert la figure de leur époch et docilement l'avenir se range à leur avis.*[28]

But anarchy and darkness supervened, painters and poets notwithstanding. A generation was decimated, Venice was bombed, and art had to choose between a plea of impotence and a confession of guilt. Yeats in his best-

* Without poets, without artists . . . everything would fall apart into chaos. There would be no more seasons, no more civilization, no more thought, no more humanity, no more life, even; and impotent darkness would reign for ever. Poets and artists together determine the features of their age, and the future meekly conforms to their edict.

known poem preserved its sovereignty by accepting its complicity in evil. Innocence has been drowned; sleep has been vexed to nightmare; and the rough beast is reborn not in spite of but because of the poet's vision. Auden likewise proclaimed the bitter triumph of art. In *Spain*, a poem inspired by the Spanish Civil War, he refuted Marxist determinism and embraced existential themes of responsibility and guilt. On the arid Peninsula the poet confronts events of his own creation, and History, no longer 'the operator, the organiser', can do no more than wring its hands in impotent pity.

This hypothesis of guilt can only be strengthened by the fate of Venice in the years of its humiliation. It is not difficult to recognize a great power and a fearful cruelty in the concerted effort of those who redeemed the city in order to mythologize its suffering. A fiction again regressed into self-fulfilling prophecy. Venice needed to be protected not only against the adversaries that art so vividly portrayed and so eloquently deplored; it needed to be protected against art itself. And History, to this supplicant as to Spain, might say Alas—but could not help nor pardon.

# EPILOGUE

THE myth required that the victim survive, because without a victim the myth itself would perish. All the modern skills of repair and restoration were therefore recruited to perfect the illusion of continuity, and soon there was no evidence of the wartime ordeal save a new chapter in the literature of tribulation. Henri de Régnier, revisiting Venice in the mid-1920s, found everything much as before. 'Je craignais', he wrote,

> après ces dix années d'absence, de retrouver une Venise autre que celle que j'avais connue et aimée et je m'aperçois qu'il n'en est rien. Son aspect est toujours le même. Les quelques dégâts qu'y a faits la guerre ont été réparés. Les œuvres d'art enlevées aux dangers qu'elles couraient ont repris leur place.*[1]

Despite the advent of *vaporetti* and modern warships, the seascape retained its age-old traffic. The gondola was spared, and for a few years yet the coloured sails of the fishing fleets moved at morning and evening in mass migration across the lagoon. The popular novelist Cecil Roberts saw them in 1922, and was inspired to write *Sails of Sunset*, a best-seller about the fishermen of Chioggia.[2] Old social habits were being resumed even before the war was over. In November 1915 *The Times* had reported: 'The *signori* have all departed; only the poor remain.'[3] By the middle of 1918, after British and Italian victories on the Piave, the *signori* were returning, and in August the bathing season was again in full swing on the Lido. 'It must be the oddest seaside place in Europe', wrote the correspondent of *The Times*,

> for the big bathing establishment . . . now lies in the midst of the Lido defences against an enemy landing. The sand is dug and revetted into trenches and a thickened barbed wire entanglement runs right across the part of the beach where the bathers bask in the sun. . . . It is odd to see this throng of people in bright bathing costumes sitting under gay sun-umbrellas, with that belt of sinister barbed wire passing amongst them, and sentries with fixed bayonets watching their water-frolics.[4]

Foreigners had been ordered to quit Venice in October 1917; but by the summer of 1919 they too were returning. They came in greater numbers

* I feared, after this ten years' absence, to find a Venice different from the one I had known and loved, but I see that this is not the case. It looks just the same. The small amount of damage inflicted by the war has been repaired. The works of art removed from the danger to which they were subjected have been put back.

than ever, hungrier than ever for fun, and richer even than before by virtue of the rapid devaluation of the lira. Before the war the rate of exchange had been 25 lire to the pound. In the early 1920s it was 100. The casino reopened and the Lido hotels were fully booked well in advance of the 1919 season. Undeterred by food-shortages, a railway strike, and political uncertainty, British tourists crowded Venice in the spring of 1920—'trying to make believe there's nothing wrong', as Horatio Brown observed.[5] The Germans and Austrians came back in droves. In the spring of 1924 Ugo Ojetti, watching them in the Piazza and hearing Viennese music coming from the cafés, found it difficult to believe that since that fateful night in 1916 when an Austrian bomb had exploded only yards from the main door of San Marco, less than eight years had elapsed. 'Mi trovo tanto solo qui stamattina a ricordare quella tetra notte', he wrote, 'che mi li sento pesare sulle spalle come fossero ottanta e fossi il sopravissuto d'un mondo svanito ormai nell'indifferente eternità.'*[6] Year by year throughout the turbulent 1920s and '30s the city intensified its orgy of pleasure. It swarmed with visitors whose lives were filled with roulette, all-day bridge, mannequin-parades, dance-bands, and noisy parties thrown by wealthy socialites—the Cunards, the Duff Coopers, the Cole Porters, the Bibescos. The marchesa Luisa Casati took over the Palazzo Dario and held there costume balls that were outrageous even by the standards of the hectic twenties. Winnie de Polignac, who resumed her entertainments for musical protégés and women friends at the Palazzo Manzoni, grew coarse and insatiable with advancing age and even this tolerant city was mildly scandalized by her Sapphic excesses. Suntan came into fashion, and every summer the Lido beach, its wartime encumbrances gone and forgotten, was crowded with denuded flesh. 'Cette plage adriatique', wrote Régnier,

> semble être devenue le lieu de campement d'une tribu sauvage composée d'individus offrant toutes les nuances du brun, de sa teinte la plus claire à sa plus sombre, les patines de tous les bronzes, la dégradation de tous les ocres, un échantillonage de tous les tons que peut revêtir la peau d'un ancien blanc et d'une ex-blanche, sous l'action des chimies solaires.†[7]

Venice was once again the scene of social encounters, the stage of Europe's carnival. The more things had changed, the more they had remained the same.

* I find myself so much alone this morning, in remembering that dreadful night, that I feel them weigh on my shoulders as if they were eighty and I were the survivor of a world now disappeared into an indifferent eternity.

† This Adriatic beach seems to have become the camping-ground of a savage tribe, composed of individuals displaying all the shades of brown, from the lightest to the darkest—the patinas of every bronze, the whole range of ochres, a sample of every tone that the skin of a once-white man or woman can acquire under the influence of solar chemistry.

Yet changes were occurring, both in the foreign response to Venice and in the foreign component of its society. The slow process of the publication of Proust's huge novel (the final sections did not appear until 1927, five years after the author's death) ensured that pre-war perceptions of Venice lingered into post-war literary France. In Britain, however, writers moved away from the Venetophile tradition, and Venice was left to historians and authors of guidebooks. Ruskin was still read by the Anglo-Saxon travelling public, and was even revered among the elderly; but he lost his influence and prestige among those of the younger generation who took themselves seriously as intellectuals. Mrs Hilbery, in Virginia Woolf's *Night and Day*, learns that 'dear Mr Ruskin' is not read by post-war educated youth. By the 1920s he had become, according to Cyril Connolly, 'the forgotten man of the nineteenth century'.[8] And no one took his place. 'Who dares write about Venice now?' asked the publisher Ernest Rhys in 1931.[9] Moreover on the rare occasions when the silence was broken there was heard a new note of harshness and hostility. An early poem of T. S. Eliot, 'Burbank with a Baedeker: Bleistein with a Cigar', evokes a Venice very different from the symphonic city that had averted the nineteenth-century incubus. Rediscovery has failed, and the Baudelairean alchemy operates in reverse, deconverting enchantment into horror. The disintoxicated vision is engulfed by the original Venice: the 'protozoic slime' that swarms with sordid existence. To the Post-Romantic, Venice signifies rich women in furs, Jewish financiers, and foreign fornicators in cheap hotels. It has become subsumed in the modern allegory of the Wasteland. Eliot's poem incorporates the old music; however, the performance is chaotic and cacophonic. Here is neither Beethoven nor Wagner, nor even Schoenberg and Stravinsky, but the desperate improvisation of amnesia. D. H. Lawrence, likewise, forgot the bewitching harmonies. He inherited the Victorians' love of Italy; yet he found it impossible to love Venice. In *Lady Chatterly's Lover* (1928) he portrayed it as a monument to Mammon, and the fleshpot of a mindless multitude.

This sudden lapsing of a habit can be attributed in part to the deflation of expression and feeling that characterized the intellectual economy between the wars. When the language of Romanticism, Decadence, and Symbolism lost currency among the highbrows, inherited valuations began to look obsolete; and Venice had been too heavily traded on the old literary market to find equivalent favour on the new. Even before Eliot and Pound imported Modernism into Anglo-Saxon letters there had been a move towards trimming, pruning, and toughening English verse. T. E. Hulme was predicting before the First World War that 'a period of hard, dry, classical verse' was coming; and Synge, Yeats, Lawrence, and the Imagists aspired to write poetry that was 'hard and clear, never blurred or indefinite'.[10] It proved very difficult to transfer Venice, already so strongly ideated within liquid, allusive writing, into such verse; and Pound's efforts to do so never transcended the limitations of self-conscious rebellion and experimentation. In prose, the style classified by Connolly as 'Mandarin' had, by the late twenties, been superseded by what he called 'New Realism'. 'Realism, simplicity, the colloquial style,' he wrote in 1938, 'they seem to have triumphed everywhere at the moment.' And however impoverished the idiom, he felt that there could be no return to the sententiousness with which the Mandarins had draped their utterance. There could be no more 'references to infinity, the remoteness of stars and planets, the littleness of man, the charm of dead civilisations, to Babylon and Troy, "on whose mouldering citadel lies the lizard like a thing of bronze"'.[11] Nor, he might have added, to things Venetian, which clearly belonged with all this lumber of worn-out diction.

Changes in literary dealing, then, explain the silence, the indifference. The hostility is perhaps more closely linked to a change in Venice itself. The city had appealed so strongly to many Victorians because it matched the elevated notion of culture that had been worked out principally by Coleridge and Matthew Arnold. After the First World War the idea of culture as the preserve of healing minorities who pursue and promote perfection came under attack, especially from the ideological left. 'Culture', wrote R. H. Tawney, 'is not an assortment of aesthetic sugar-plums for fastidious palates, but an energy of the soul. . . . When it feeds on itself, instead of drawing nourishment from the common life of mankind, it ceases to grow; and when it ceases to grow, it ceases to live.'[12] Nevertheless in much critical writing the old hierarchies of class and taste lived on. T. S. Eliot, F. R. Leavis, and I. A. Richards were still working out variations on the Coleridgean–Arnoldian theme of culture as the superfine attribute of a privileged class; and it may be that Venice fell out of favour with this generation of intellectuals because it failed, in the age of mass tourism, to conform to the traditional prescriptions of 'cultivation' and suggested more strongly the undesirable features of 'civilization'. After the war, foreign society in Venice seemed, to use Coleridge's terms, more varnished than polished. Soon there were not many left of the old resident Venetophiles and connoisseurs. A conspicuous absence was that of Fritz von Hohenlohe, whose Austrian nationality had meant expulsion from Italy in the spring of 1915. Vigorously abjuring his country and denouncing his compatriots, he spent the war years unhappily in Switzerland, and died in Rapallo in 1923.[13] And Venice lost its Anglo-Saxon colony—that leisured circle of expatriates who had nurtured predilections for fine art and fine living in the manner of a Coleridgean clerisy.

Horatio Brown told Rosebery in June 1919 that 'a good many of the old colony' had returned to Venice, 'but', he added, 'we keep pretty quiet'.[14] Constance Fletcher, who had remained in the city as a nurse throughout the war, was, like Horatio, soon forced by poverty to move to smaller premises. She gave up the Palazzo Cappello and took an apartment on the Zattere, where she died in 1938. Among those who came back with Horatio after the war were the artists Clara Montalba and Henry Woods, and the pastor Alexander Robertson. Robertson, obscurely immortalized as a fragment in the mosaic of Ezra Pound's Canto LXXVI, had been Church of Scotland chaplain in Venice since 1888. He now became an enthusiastic convert to Fascism, wrote a eulogistic biography of Mussolini, and was decorated by the Italian Crown. But to most surviving members of the old colony post-war Venice was less congenial, and they died elsewhere—Mrs Curtis in New York; Lady Radnor at Ascot; Althea Wiel and Caroline Eden in London. Horatio would have ended his days in Scotland had he not been too ill to travel. They had no successors, and the institutions they had patronized vanished with them. The Ospedale Cosmopolitano closed in 1933; the Sailors' Institute a few years later. St George's was closed in 1936, and did not reopen until 1952. Laura Ragg, looking back in 1936, made a sad note of the change. 'Today,' she wrote, 'while Venice attracts a cosmopolitan herd of tourists and bathers, and two or three of its English residents cling to it, it cannot be said to have an Anglo-American colony.' Mrs Ragg, like most of the other survivors of that colony, ended her life far from Venice, and even farther from what Henry James had called 'that incredible past in which we once lived . . . not knowing that we were fantastically happy'. She died in Bath, in 1962, in her ninety-seventh year.[15]

# NOTES

Throughout the notes the following abbreviations are used:

| | |
|---|---|
| BL | British Library |
| Bodleian | Bodleian Library, Oxford |
| Brotherton | Brotherton Collection, University of Leeds |
| BUL | Bristol University Library |
| HLRO | House of Lords Record Office |
| NLS | National Library of Scotland |
| SCL | Sheffield Central Library |

The place of publication of printed works is omitted where this is London. Full names and titles are given in initial references; thereafter abbreviations are used.

*Introduction*

1. Jules and Edmond de Goncourt, *Journal: Mémoires de la vie littéraire* (Paris, 1956), ii. 320–1.
2. René Pierre Colin, *Schopenhauer en France* (Lyons, 1979), 149–207.
3. Arthur Symons, *Cities of Italy* (1907), 75–6.
4. Ugo Ojetti, *Cose viste*, new edn. (Florence, 1951), i. 178.
5. Gabriel Monod, *Les Maîtres de l'histoire*, 5th edn. (Paris, n.d.), 41.
6. Quoted in Carl Becker, *The Heavenly City of the Eighteenth-Century Philosophers* (New Haven, Conn., 1932), 94. This section is based on Becker's book and on Duncan Forbes's *The Liberal Anglican Idea of History* (Cambridge, 1952).
7. Quoted by Forbes, *Liberal Anglican Idea*, 133.
8. Quoted by Colin, *Schopenhauer en France*, 152.
9. J. A. Froude, *Thomas Carlyle*, ed. John Clubbe (1979), 418.
10. Friedrich Nietzsche, *The Birth of Tragedy*, trans. Francis Golffing (New York, 1956), ch. 15.
11. Arthur Schopenhauer, *The World as Will and Representation*, trans. E. F. J. Payne (1969), i. sect. 52.
12. Hector Berlioz, *Mémoires*, new edn. (Paris, 1969), i. 137–40; Richard Wagner, *Beethoven*, trans. E. Dannreuther (1903), *passim*.
13. Théophile Gautier, *Voyage en Italie*, new edn. (Cambridge, 1904), 24–5.
14. John Ruskin, *The Stones of Venice*, Everyman edn. (London, 1907), i. ch. 1, sects. 24, 25.
15. Jules and Edmond de Goncourt, *Préfaces et manifestes littéraires*, new edn. (Paris, n.d.), 48.
16. Ian Watt, *The Rise of the Novel*, new edn. (Harmondsworth, 1963), ch. 1.
17. Quoted in Philip Collins, 'Dickens and London', in H. J. Dyos and Michael Wolff (eds.), *The Victorian City* (1973), ii. 541.
18. Goncourt, *Préfaces*, 73, 163.

19. Goncourt, *Journal*, ii. 96.
20. Henry James, *The House of Fiction*, ed. Leon Edel, new edn. (1962), 25, 57, 105–6, 170, 173, 241, 245–6.
21. Quoted, from Flaubert's *La Première éducation sentimentale* and Maupassant's *Étude sur Gustave Flaubert*, in Colin, *Schopenhauer en France*, 152–3.
22. James, *House of Fiction*, 134.
23. Henry James, *Italian Hours* (1909), 73.
24. Matthew Arnold, *Culture and Anarchy*, 3rd edn. (1882), preface. The classic treatment of the subject is Raymond Williams's *Culture and Society* (Harmondsworth, 1961).
25. J. A. Symonds, *In the Key of Blue* (1893), 195–215.
26. George Eliot, *Felix Holt*, new edn. (Harmondsworth, 1972), 621.
27. Samuel Taylor Coleridge, *On the Constitution of the Church and the State*, ed. John Colmer (Princeton, NJ, 1976), 42–9; T. S. Eliot, *Selected Essays* (1951), 431–43.
28. See Sand's *Mattea* (1842), Barrès's *L'Ennemi des lois* (1893), Rolfe's *The Desire and Pursuit of the Whole* (1934), and Powys's *Autobiography* (1934). For a discussion of the whole issue of androgyny see A. J. L. Busst, 'The Image of the Androgyne in the Nineteenth Century': Ian Fletcher (ed.), *Romantic Mythologies* (1967), 12–76.
29. See especially Giandomenico Romanelli, *Venezia ottocento* (Rome, 1977); Giuseppe Pavanello and Giandomenico Romanelli (eds.), *Venezia nell'ottocento: Immagini e mito* (Milan, 1983). The second item is the catalogue of an exhibition held in Venice 1983–4.

*Chapter I*

1. Kent Robert Greenfield, 'Commerce and Enterprise at Venice 1830–1848', *Journal of Modern History*, 11 (Chicago, 1939), 313–33; Alvise Zorzi, *Venezia austriaca* (Rome, 1986), 109–10, 135; G. M. Trevelyan, *Manin and the Venetian Revolution of 1848* (1923), 45–7; Piers Brendon, *Thomas Cook & Son: 150 Years of Popular Travel* (1991), 112; Edmund Flagg, *The City of the Sea* (1853), 55; Daniel Pidgeon, *Venice* (1895), 28; George Augustus Sala, *Rome and Venice* (1869), 44–5; Henry James, *Italian Hours* (1909), 7, 28; Katherine Furse, *Hearts and Pomegranates* (1940), 50–1; Lonsdale and Laura Ragg, *Things Seen in Venice* (1912), 170; *The Times*, 3 Aug. 1886, 3 Sept. 1918; Edward Hutton, *Venice and Venetia* (1911), 178; John Murray (pub.), *Handbook for Travellers in Northern Italy*, 16th edn. (1897), 328; Frederick Rolfe, *The Venice Letters*, new edn. (1987), 76.
2. Ezra Pound and Dorothy Shakespeare, *Ezra Pound and Dorothy Shakespeare, Their Letters*, ed. O. Pound and A. W. Litz (1985), 212, 219. See also Helen Matilda Bouverie (Lady Radnor), *From a Great-Grandmother's Armchair* (1927), and Augusta Gregory, *Seventy Years* (Gerrards Cross, 1974).
3. *Venice Mail*, 31 Jan. 1874; The Venice Sailors' Institute, *Annual Report for 1890* (Venice, 1891); *The Times*, 11 Oct. 1894; L. V. Fildes, *Luke Fildes, RA* (1968), 84; Bouverie, *From a Great-Grandmother's Armchair*, 190; Henry James, *The Letters of Henry James*, ed. Leon Edel (Cambridge, Mass., 1974–84), iii. 475;

Henry James, *The Princess Casamassima*, ch. 30; *The Wings of the Dove*, bk. 8, ch. 2, bk. 9, ch. 2; *The Aspern Papers*, ch. 5; Camillo Boito, *Storielle Vane*, new edn. (Florence, 1970), 436.

4. Flagg, *City of the Sea*, 64.
5. H. Thirria, *La Duchesse de Berry* (Paris, 1900), 347, 412–20; Mary Lutyens (ed.), *Effie in Venice* (1965), 90, 91, 107; W. D. Howells, *A Fearful Responsibility*, new edn. (Edinburgh, 1882), 15; C. E. Norton, 'Rawdon Brown and the Tomb of Norfolk', *Atlantic Monthly*, June 1889; C. E. Norton, *The Letters of Charles Eliot Norton*, ed. Sarah Norton and M. A. D. Howe (1913), i. 405–6; Paul Kaufman, 'Rawdon Brown and his Adventures in Venetian Archives', *English Miscellany*, 18 (1967), 283–303; E[lizabeth] E[astlake], 'Rawdon Brown', *The Times*, 8 Sept. 1883.
6. Viola Bankes, *A Dorset Heritage* (1953); The National Trust, *Kingston Lacy, Dorset* (1988); A. L. Rowse, 'Byron's Friend Bankes', *Encounter*, 44 (1975), 25–32.
7. Edward Leeves, *Leaves from a Victorian Diary*, ed. John Sparrow (1985).
8. S. M. Ellis, *The Solitary Horseman: The Life and Adventures of G. P. R. James* (1927), 236–7; *The Times*, 19 Mar. 1896. Seventeen British graves were relocated on this occasion, including that of James.
9. Edwin Lee, *Bradshaw's Invalid's Companion to the Continent*, 2nd edn. (1861), 94; T. H. Burgess, *The Climate of Italy in Relation to Pulmonary Consumption* (1852), 102–21; R. E. Scoresby-Jackson, *Medical Climatology* (1862), 408–10.
10. Robert Browning, *Learned Lady: Letters to Mrs Thomas Fitzgerald*, ed. Edward McAleer (1966), 167, 182; *Builder*, 19 June 1886; *Building News*, 18 June 1880; *The Times*, 7 Aug. 1886; *Lancet*, 28 Aug. 1886; Guy de Maupassant, 'Venise 1885', *La Vie errante* (Paris, 1926); Anne Buckland, *The World Beyond the Esterelles* (1884), i. 210; Charles Dickens, *Little Dorrit*, bk. 2, ch. 6; Journal of Enid Layard: BL Add. MSS 46162, fos. 102, 123.
11. Honoré de Balzac, *Lettres à Mme Hanska*, ed. Roger Pierrot (Paris, 1967), i. 489; George Sand, *Œuvres autobiographiques* (Paris, 1971), ii. 217; Lutyens (ed.), *Effie in Venice*, 91 n.
12. Mary Shelley, *Rambles in Germany and Italy* (1844), ii. 123.
13. Nassau Senior, *Journals Kept in France and Italy*, 2nd edn. (1871), i. 345; Anon. 'Venice', *Quarterly Review*, 86 (1849–50), 184–227; Ellis, *Solitary Horseman*, 245; W. D. Howells, *Venetian Life*, new edn. (1883), i. 115; John Ruskin, *Letters from Venice*, ed. J. C. Bradley (New Haven, Conn., 1955), 5.
14. Sand, *Œuvres autobiographiques*, ii. 211.
15. Ruskin, *The Stones of Venice*, iii. app. 3.
16. O. A. Sherrard, *Two Victorian Girls* (1966), 272–3. See also Howells, *Venetian Life*, i. 24.
17. Senior, *Journals*, i. 345; John Ruskin, *Reflections of a Friendship: The Letters of John Ruskin to Pauline Trevelyan*, ed. Virginia Surtees (1979), 23–4; Lutyens (ed.), *Effie in Venice*, 91.
18. Sala, *Rome and Venice*, 232. See also *The Times*, 8 Aug. 1860, 30 Aug. 1860, 4 Sept. 1860.
19. Mrs Newman Hall, *Through the Tyrol to Venice* (1860), 277–8; H. Taine,

*Voyage en Italie*, 5th edn. (Paris, 1884), ii. 295; Diary of Georgina Max Muller: Bodleian MSS Dep/d/199; Howells, *Venetian Life*, i. 9–10, 16–23; ii. 252; Richard Wagner, *My Life* (1911), ii. 696.

20. Zina Hulton, 'Fifty Years in Venice', unclassified MSS, Bodleian Library; Browning, *Learned Lady*; Gregory, *Seventy Years*.
21. Laura Ragg, 'Venice when the Century Began', *Cornhill Magazine*, Jan. 1936.
22. Paolo Cossato, 'La Carriera di un teatro d'opera', in T. Pignatti (ed.), *Gran teatro La Fenice* (Venice, 1981); Gabriel Fauré, *Correspondance*, ed. J. M. Nectoux (Paris, 1980), 71.
23. Rudolfo Pallucchini, 'Significato e valore della "Biennale" nella vita artistica veneziana e italiana', in Vittore Branca (ed.), *Storia della civiltà veneziana* (Florence, 1979), iii; William Rossetti, 'British Art in Venice': *Athenaeum*, Jan. 1909.
24. Fildes, *Luke Fildes*, 70, 178; Hulton, 'Fifty Years in Venice'; K. Baedeker, *Handbook for Travellers to Northern Italy*, 10th edn. (Leipzig, 1895), 234; Romain Rolland, *Retour au palais Farnèse* (Paris, 1956), 346.
25. Robert Browning, *The Letters of Robert Browning Collected by Thomas J. Wise*, ed. T. L. Hood (1933), 241, 250, 252, 266; Mrs Sutherland Orr, *The Life and Letters of Robert Browning* (1891), 359–62; Browning, *Learned Lady*, 189–90; Daniel Sargent Curtis, 'Robert Browning, 1879 to 1885', in Robert Browning, *More Than Friend: The Letters of Robert Browning to Katharine de Kay Bronson*, ed. Michael Meredith (Waco, Tex., 1985); Katharine de Kay Bronson, 'Browning in Venice': *Cornhill Magazine*, 12 (1902), 149–71; Princesse Edmond de Polignac, 'Memoirs': *Horizon*, 12 (1945), 110–41; Philip Kelley and Betty A. Coley, *The Browning Collections* (Waco, Tex., 1984), pp. xix, xx.
26. Hulton, 'Fifty Years in Venice'; Howells, *Venetian Life*, ii. 152; Horatio Brown, *In and Around Venice* (1905), 201; J. A. Banks, *Prosperity and Parenthood* (1965), 93, 204.
27. Frederick Eden, *A Garden in Venice* (1903), 87; J. A. Menzies, 'The Cholera Epidemic in Venice', *Lancet*, 28 Aug. 1886; Ragg, *Things Seen in Venice*, 247.
28. Stuart Tidey, 'Travel in Europe', in Edmund Hobhouse (ed.), *Health Abroad: A Medical Handbook of Travel* (1899), 276–7.

*Chapter II*

1. Browning, *Learned Lady*, 173.
2. J. A. Symonds, *The Letters of John Addington Symonds*, ed. Herbert Schueller and Robert Peters (Detroit, 1967–9), iii. 516, 661, 662.
3. Princess Victoria (Empress Frederick of Germany), *The Empress Frederick Writes to Sophie: Letters 1889–1901*, ed. Arthur Gould Lee (1955), 124–6; J. A. Symonds, *Letters*, iii. 763, 771; Hulton, 'Fifty Years in Venice'.
4. The quotation is from 'Casa Alvisi', in *Italian Hours*.
5. Nora Wydenbruck, *Rilke, Man and Poet* (1946), 216–21; R. M. Rilke, *Selected Letters*, trans. R. F. Hull (1946), 301; Gino Damerini, *D'Annunzio e Venezia* (Verona, 1943), 53–4.
6. Henri Martineau, *P. J. Toulet, Jean de Tinan, et Madame Bulteau* (Paris, 1958),

20–71; Henri de Régnier, *L'Altana ou la vie vénétienne 1899–1924*, 10th edn. (Paris, 1928), i. 62; Polignac, 'Memoirs'; Damerini, *D'Annunzio e Venezia*, 161.

7. Polignac, 'Memoirs'; Michael de Cossart, *The Food of Love: The Princesse Edmond de Polignac and Her Circle* (1978); *The Times*, 24 Aug. 1887; Fauré, *Correspondance*, 154–7.
8. *The Times*, 30 Apr. 1888; James, *Italian Hours*, 69.
9. Gregory, *Seventy Years*, 442, 444.
10. James, *Letters*, iv. 426–7; *The Times*, 11 June 1938; Mabel Dodge Luhan, *European Experiences* (New York, 1935), 290–9.
11. John Cowper Powys, *Autobiography* (1934), 411; Victor Hall, 'The Last Years: Some Memories of Rolfe in Venice Recalled by Mrs Ivy van Someren', in Cecil Woolfe and Brocard Sewell (eds.), *Corvo, 1860–1960* (Aylesford, 1961). On Rolfe generally see A. J. A. Symons, *The Quest for Corvo* (1934) and Miriam J. Benkovitz, *Frederick Rolfe, Baron Corvo* (1977). *The Desire and Pursuit of the Whole* was finally published, slightly bowdlerized, in 1934.
12. *The Times*, 30 Apr. 1888, 1 Oct. 1889; The Additional Curates Society, *St George's Venice* (Venice, n.d.); Journal of Enid Layard: BL Add. MSS 46163, fo. 49. For Layard see Gordon Waterfield, *Layard of Nineveh* (1963); Gregory, *Seventy Years*; and John Fleming, 'Art Dealing and the Risorgimento', *Burlington Magazine*, 115 (1973), 4–16.
13. Hulton, 'Fifty Years in Venice'; Ragg, 'Venice when the Century Began'; Lina Waterfield, *Castle in Italy* (1961), 68–70; Julia Cartwright, *A Bright Remembrance*, ed. Angela Emanuel (1989), 277, 284–6, 295, 304; Bouverie, *From a Great-Grandmother's Armchair*, 190; *The Times*, 29 Sept. 1903, 9 Nov. 1912, 17 Jan. 1913; Journal of Enid Layard, BL Add. MSS 46168, fos. 12, 44, 149–50, 152, 155–6, 159, 190; ibid. 46169, fos. 3, 96, 261; ibid. 46170, fos. 34, 43–4; Frederick Rolfe, *The Desire and Pursuit of the Whole* (1934), 54–5, 93, 110, 123, 144, 182; Cecil Woolf, editorial intro. to Baron Corvo, *The Venice Letters* (1987); Benkovitz, *Frederick Rolfe*, 269; Frederick Rolfe, *Letters to C. H. C. Pirie-Gordon*, ed. Cecil Woolf (1959), 90; J. A. Symonds, *Letters*, iii. 762–3. See also Waterfield, *Layard of Nineveh*, and Gregory, *Seventy Years*.
14. Eden, *A Garden in Venice*; Hulton, 'Fifty Years in Venice'; Rilke, *Selected Letters*, 303–4; Ragg, 'Venice When the Century Began'; Anna de Noailles, *La Domination* (Paris, 1905); Henri de Régnier, *La Peur de l'amour* (Paris, 1907).
15. James, *Letters*, iii. 287, 437; Henry James, *William Wetmore Storey and His Friends* (1903), 285; Gregory, *Seventy Years*, 172, 200; Pidgeon, *Venice*, 7–8; Orr, *Robert Browning*, 360; Robert Browning, *Letters . . . Collected by T. J. Wise*, 239–41, 247, 266; Robert Browning, *New Letters of Robert Browning*, ed. W. C. de Vane and K. L. Knickerbocker (1951), 363; Browning, *Learned Lady*, 71, 104–6; Bronson, 'Browning in Venice'; Kelley and Coley, *Browning Collections*, pp. ix–xii, 536–43, 608–10; Oscar Browning, *Memories of Sixty Years* (1910), 6; Lord Ronald Sutherland Gower, *Old Diaries, 1881–1901* (1902), 149; Lillian Whiting, *The Brownings: Their Life and Art* (1911), 262–4; Katherine Bradley and Edith Cooper ('Michael Field'), *Work and Days*, ed. T. and D. C. Sturge-Moore (1933), 206–21; William Lyon Phelps, *Autobiography with Letters* (New York, 1939), 448–56; Maisie Ward, *The Tragi-Comedy of Pen Browning* (New York, 1972); Meredith, editorial intro. to Browning, *More than*

*Friend*; Luhan, *European Experiences*, 115–20; A. H. Sayce, *Reminiscences* (1923), 180; Alfred Domett, *Diary*, ed. E. A. Horseman (1953), 55–6, 169, 179–215, 242, 248; Betty Miller, *Robert Browning* (1952), 280; Régnier, *L'Altana*, ii. 143; Journal of Enid Layard, BL Add. MSS 46163, fo. 39; ibid. 46164, fo. 176; ibid. 46166, fos. 85–185.

16. James, *Letters*, iii. 166; iv. 182–3; Henry James, *The Complete Notebooks*, ed. Leon Edel and Lyall Powers (New York, 1987), 221–2; James, *Italian Hours*, 79–81; Meredith, editorial intro. to Browning, *More Than Friend*; Browning, *Learned Lady*, 129, 187; Bronson, 'Browning in Venice'; Journal of Enid Layard, BL Add. MSS 46167, fo. 55. Enid Layard recorded Mrs Moore's confidential revelation on 27 July 1885. See ibid 46162.
17. James, *Letters*, iii. 195–6, 287, 389–90; iv. 451–2; James Rennell Rodd, *Social and Diplomatic Memories, Second Series 1894–1901* (1923), 44–5; Stanley Olsen, *John Singer Sargent* (1986), 84–6; Cartwright, *A Bright Remembrance*, 288; J. A. Symonds, *The Letters and Papers of John Addington Symonds*, ed. H. F. Brown (1923), 210–11; Evelyn Barclay, MS Diary in the Armstrong Browning Library, Baylor University, Texas; Robert Browning, *Learned Lady*, 127–8; Curtis, 'Robert Browning, 1879–1885'.
18. Hulton, 'Fifty Years in Venice'; Gower, *Old Diaries*, 229; Ragg, 'Venice when the Century Began'.
19. Margaret Symonds, *Days Spent on a Doge's Farm* (1893); Symonds, *Letters*, iii. 524–5; Henry James, *Letters*, iii. 170–1; Gregory, *Seventy Years*, 287; Ellis, *The Solitary Horseman*, 242; Louise Hall Tharp, *Mrs Jack: A Biography of Isabella Gardner* (Boston, Mass., 1965), 147–8.
20. Bouverie, *From a Great-Grandmother's Armchair*, 222.
21. Ragg, 'Venice when the Century Began'.
22. Ragg, *Things Seen in Venice*, 53.
23. Horatio F. Brown, *In and Around Venice*, 200–1.
24. Rolfe, *Desire and Pursuit*, 125.
25. Bouverie, *From a Great-Grandmother's Armchair*.
26. Gower, *Old Diaries*, 156; Symonds, *Letters*, vols. ii, iii; Margaret Symonds, *Out of the Past* (1925), 240; Furse, *Hearts and Pomegranates*, 46, 197; Symonds family correspondence: BUL MS DM/376.
27. Quoted in Geoffrey Faber, *Jowett* (1957), 370.
28. David Newsome, *On the Edge of Paradise: A. C. Benson the Diarist* (1980), 286.
29. Symonds, *Letters*, iii. 754; Diary of Mr Hill of Edinburgh, Bodleian MS top. gen. e. 77; Rolfe, *Venice Letters*, 32–3; Laura Ragg, 'Epilogue', in Frederick Rolfe, *Letters to R. M. Dawkins*, ed. Cecil Woolf (1962).
30. Quoted by Ragg: 'Venice when the Century Began'.

## *Chapter III*

1. James, *Complete Notebooks*, 33.
2. Preface to the New York edn. of *The Aspern Papers*.
3. James, *Italian Hours*, 69, 82.
4. James, *The House of Fiction*, 38.
5. Hulton, 'Fifty Years in Venice', fo. 102.

6. Luhan, *European Experiences*, 293–4.
7. Leon Edel, *Henry James: The Treacherous Years* (1969), 119.
8. James, *Letters*, iii. 398.
9. Edel, *Henry James: The Treacherous Years*, 119.
10. Rolfe, *Desire and Pursuit*, 231.
11. Pound and Shakespeare, *Letters*, 224; Leonard Woolf, *Beginning Again* (1964), 72.
12. Strachey Papers, HLRO MS S/33/4/25.
13. Rosebery Papers, NLS MS 10122.
14. Gosse Papers, Brotherton.
15. Rosebery Papers, NLS MS 10127. Some of the letters received by Symonds, including those from Gosse, Edward Carpenter, and Henry Sidgwick, were returned to their senders. The letters written by Symonds to Sidgwick were subsequently acquired by Brown—presumably on Sidgwick's death in 1900.
16. Virginia Woolf, 'The Art of Biography': *Collected Essays*, 4 (1967).
17. Harold Nicolson, *The Development of English Biography* (1927), 145.
18. Symonds, *Letters*, i. 846.
19. Ibid. iii. 450.
20. Ibid. iii. 839.
21. Gosse Papers, Brotherton.
22. BUL MS DM/188.
23. Symonds, *Letters*, iii. 840.
24. C. J. Holmes, *Self and Partners* (1936), 151–2.
25. Carpenter Papers, SCL MS 386/53.
26. Letter to Carpenter, 22 July 1893: Carpenter Papers, SCL MS 386/43.
27. See letter to Carpenter, 14 Feb. 1895: Carpenter Papers, SCL MS 386/52.
28. BUL MS DM/376. See also Phyllis Grosskurth, *Havelock Ellis* (1981), 181.
29. Grosskurth, *Havelock Ellis*, 181–2.
30. Carpenter Papers, SCL MS 386/76.
31. Grosskurth, *Havelock Ellis*, 193–201.
32. V. Woolf, 'The Art of Biography'.
33. BUL MS DM/367.
34. Strachey Papers, HLRO MS S/33/4/31.
35. Rosebery Papers, NLS MS 10124.
36. Ibid. 10126.
37. Ragg, 'Venice when the Century Began'.
38. L. Woolf, *Beginning Again*, 72.
39. Letter of 31 Oct. 1923: BUL MS DM/367.
40. Furse, *Hearts and Pomegranates*, 80.
41. Letter from Catherine Symonds to Katherine Symonds: BUL DM/1279, 'Furse Correspondence', Box 1.
42. Nigel Nicolson and Joanne Trautman (eds.), *The Letters of Virginia Woolf* (1972–7), i. 318.
43. *Dictionary of National Biography*.
44. Frances Spalding, *Vanessa Bell* (1983), 181–2.
45. Virginia Woolf, *The Diary of Virginia Woolf*, ed. Anne Oliver Bell (1977–84), ii. 114, 121–2.
46. L. Woolf, *Beginning Again*, 73.

47. Symonds, *Letters*, iii. 419, 711.
48. Letter of 24 Apr. 1922, BUL MS DM/367.
49. BUL MS DM/367.
50. Symonds, *Letters and Papers*, p. vi. This volume contains 250 letters, chosen out of the 2,500 in Brown's possession. Of these, over 2,000 were addressed to himself, and some 400 were addressed to Henry Sidgwick.
51. BUL MS DM/367.
52. Letters to Rosebery of 26 Mar. 1923; 13 Apr. 1923; 7 May 1923, NLS MS 10127.
53. V. Woolf, *Diary*, ii. 121.
54. Margaret Symonds, *Out of the Past* (1925), pp. ix–x.
55. Rosebery Papers, NLS MS 10127.
56. It seems possible that these papers included material on sexual inversion. On 31 Mar. 1925 Horatio wrote to Madge: 'You will recollect that one of the correspondents, [F. W.] Myers, in the batch you recently sent me, thought the letters so private that he *volunteers* to your father to guarantee their anonymity. "Henry Sidgwick", he writes, "and I agree that to no one except Gurney would we make known *even the initials* of those concerned"'. BUL MS DM/367.
57. BUL MS DM/367.
58. Ibid.
59. Ibid.
60. Brown's will is in the Scottish Record Office at SC/70/4/605.
61. Memorandum by Katharine Furse, 27 Apr. 1927: BUL MS DM/911.
62. Janet Vaughan to the editors of Symonds's letters, *The Letters of J. A. Symonds*, ii. 381 n. Dame Janet's account is corroborated by information given to Katharine Furse by Hagberg Wright in 1927—see Furse's 'Memorandum', BUL MS DM/911. In 1939 Wright told Furse that he had no recollection of Gosse's involvement, but this was the year before his death and he was probably senile. It is noteworthy that Gosse did not destroy the letters from Symonds among his own papers, nor were they destroyed after his death in 1928. They survive in the Gosse archive, in the Brotherton Collection, University of Leeds.
63. Havelock Ellis, *Studies in the Psychology of Sex*, i (New York, 1936), p. xi.
64. BUL MS DM/911.
65. Letter of 26 Mar. 1923, Rosebery Papers, NLS MS 10127.
66. BUL MS DM/198/1a.
67. Ibid.
68. BUL MS DM/911.
69. BUL MS DM/190/6.
70. BUL MS DM/911; DM/198/1a.
71. BUL DM/911.
72. BUL DM/190/6.

*Chapter IV*

1. J. A. Froude, *Short Studies on Great Subjects* (London and Glasgow, n.d.), 28.
2. Leopold Ranke, *History of England, Chiefly in the Seventeenth Century*, Eng.

trans. (Oxford, 1875), v. 427–8; Leopold Ranke, *History of the Reformation in Germany*, trans. Sarah Austin, new edn. (1905), p. ix.

3. Gertrude Himmelfarb, *Lord Acton: A Study in Conscience and Politics* (Chicago, 1952), 202; Lord Acton, *Lectures on Modern History*, new edn. (1920), 5, 7; Damian McElrath (ed.), *Lord Acton: The Decisive Decade* (Louvain, 1970), 39–40.
4. Horatio F. Brown, *Studies in the History of Venice* (1907), ii. 150–1.
5. Horatio F. Brown, *Venetian Studies* (1887), 178.
6. Rawdon Brown, *Preface to the Calendar of State Papers and Manuscripts Relating to English Affairs Existing in the Archives and Collections of Venice and other Libraries in North Italy* (1864), pp. v, xxviii–xxxii.
7. Ibid. p. v; H. Brown, *Venetian Studies*, 211–12; Sebastian Giustinian, *Four Years at the Court of Henry VIII*, ed. Rawdon Brown (1854), i. pp. xi–xii; Armand Baschet, *La Diplomatie vénitienne* (Paris, 1862), 40–58; Charles Yriarte, *La Vie d'un patricien de Venise au* XVI[e] *siècle* (Paris, 1884), 93.
8. Theodore H. von Laue, *Leopold Ranke: The Formative Years* (Princeton, NJ, 1950), 35; Baschet, *Diplomatie vénitienne*, 1; Armand Baschet, *Les Archives de Venise* (Paris, 1857), pp. xiv–xv.
9. Acton, *Lectures on Modern History*, 315.
10. Ranke, *Reformation in Germany*, p. xi.
11. J. R. Seeley, *The Expansion of England*, new edn. (1909), 201.
12. Laue, *Ranke*, 34 n.; Leonard Krieger, *Ranke: The Meaning of History* (Chicago, 1977), 105, 117–19; Baschet, *Diplomatie vénitienne*, 69; Lord Acton, *Essays in the Study and Writing of History*, ed. J. R. Fears (Indianapolis, 1985), 167; Acton, *Lectures*, 7; Leopold Ranke, *The Ottoman and Spanish Empires in the Sixteenth and Seventeenth Centuries*, trans. W. K. Kelley (1843), 1; Ranke, *History of England*, i. p. xiii; Leopold Ranke, *Civil Wars and Monarchy in France in the Sixteenth and Seventeenth Centuries*, trans. M. A. Garvey (1852), i. p. vii.
13. Krieger, *Ranke*, 117; Ranke, *Ottoman and Spanish Empires*, 1–4; Leopold Ranke, *The History of the Popes*, trans. E. Foster (1847), i. pp. xii–xiii; Gabriel Monod, *Les Maîtres de l'histoire: Renan, Taine, Michelet*, 5th edn. (Paris, n.d.), 137–9.
14. M. L. de Mas Latrie, *Histoire de l'isle de Chypre sous le règne des princes de Lusignan* (Paris, 1852–5), ii. pp. iii, iv, x; McElrath (ed.), *Acton*, 53, 122–30; Van Wyck Brooks, *New England Summer* (1941), 38; BL Add. MSS 38990, fos. 322, 343, 355; Howells, *A Fearful Responsibility*, 14.
15. Baschet, *Diplomatie vénitienne*, 1–4, 23–34, 80–5.
16. Eugenio Alberi (ed)., *Relazioni degli ambasciatori veneti al Senato*, 1st Ser. (Florence, 1839–63), i. pp. vii–xiii.
17. R. Brown, *Preface*, p. lxxiii.
18. Kaufman, 'Rawdon Brown and his Adventures in Venetian Archives'.
19. BL Add. MSS 39043, fo. 185.
20. Ibid. 39100, fo. 311.
21. Acton, *Essays*, 165; Lord Acton, *Historical Essays and Studies* (1919), 353; Acton, *Lectures*, 7; G. P. Gooch, *History and Historians in the Nineteenth Century* (1913), 94.

22. McElrath (ed.), *Acton*, 129; H. Brown, *Venetian Studies*, 211; Herbert Butterfield, *Man on his Past*, new edn. (1969), 90; Lord Acton, *Letters of Lord Acton to Mary Gladstone*, 2nd edn. (1913), 179; Lord Acton, *The Correspondence of Lord Acton and Richard Simpson*, ed. J. L. Althoz, D. McElrath, and J. C. Holland (Cambridge, 1973), ii. 25; Acton, *Essays*, 167; Lionel Kochan, *Acton on History* (1954), 130–1. Gindely's comments are cited by Gooch, *History and Historians*, 92 n.
23. Gooch, *History and Historians*, 94; J. B. Bury, *Selected Essays* (Cambridge, 1930), 19–20; Karl Lamprecht, *What is History?* (New York, 1905), 25.
24. Kochan, *Acton on History*, 20; Acton, *Essays*, 167; Acton, *Lectures*, 15; Acton, *Letters to Mary Gladstone*, 178.
25. Acton, *Historical Essays and Studies*, 353, 393, app.; Acton, *Letters to Mary Gladstone*, 46, 96; Acton, *Essays*, 440; Acton, *Lectures*, 315–17; Butterfield, *Man on his Past*, 85.
26. James Bryce, *Studies in Contemporary Biography* (1911), 391–2.
27. C. A. Sainte-Beuve, *Causeries du lundi*, 3rd edn. (Paris, n.d.), ix. 439.
28. Acton, *Historical Essays and Studies*, 393.
29. Acton, *Essays*, 459.

*Chapter V*

1. Henry Hallam, *A View of the State of Europe in the Middle Ages*, 7th edn. (1837), i. 483; James Fenimore Cooper, *The Bravo*, new edn. (New York, 1963), 147, 382; Edward Smedley, *Sketches from Venetian History* (1831), ii. 96; John Edward Bowden, *The Life and Letters of Frederick William Faber* (1869), 107; Ruskin, *The Stones of Venice*, i. ch. 1, sects. 8, 9.
2. Guy Dumas, *La Fin de la république de Venise* (Paris, 1964), 18–19.
3. J. C. L. Sismonde de Sismondi, *Histoire des républiques italiennes du moyen âge*, 2nd edn. (Paris, 1818), esp. iv, viii, ix.
4. For Daru see Sainte-Beuve, *Causeries du lundi*, ix. 430–69; Stendhal, *La Vie de Henri Brulard*, new edn. (Paris, 1961); *Dictionnaire biographique française*.
5. Pierre Daru, *Histoire de la république de Venise*, 3rd edn. (Paris, 1826), i. 4, 84; iii. 209; vi. 207; vii. 339–72.
6. Ibid. iii. 144–7; vii. 282–330; viii. 1–103.
7. Ibid. v. 175.
8. V. del Litto, 'Stendhal et Venise', in Carlo Pellegrini (ed.), *Venezia nelle letterature moderne* (Venice, 1961), 99–106.
9. *Quarterly Review*, 31 (1824–5), 427, 443.
10. Ugo Foscolo, *Scritti vari di critica storica e letteraria*, ed. Uberto Limentani (Florence, 1978), pp. lxxvii, 572–4, 661–2.
11. Domenico Tiepolo, *Discorsi sulla storia veneta: cioè rettificazione di alcuni equivoci nella storia del Daru* (Udine, 1828). Reprinted in Daru, *Histoire*, 4th edn. (Paris, 1853), ix.
12. Dumas, *La Fin de la république de Venise*, 554–6.
13. Léon Galibert, *Histoire de la république de Venise*, new edn. (Paris, 1850), 3, 461.
14. Jules Michelet, *Histoire de France*, new edn. (Paris, n.d.), ix. 250–1.

15. Yriarte, *La Vie d'un patricien*, 259–60, 348–9; Baschet, *La Diplomatie vénétienne*, 5–6. Baschet accused Daru of deliberate falsification—see *les Archives de Venise*, 12.
16. Vladimir Lamansky, *Secrets d'état de Venise* (St Petersburg, 1884), pp. vii–viii, 761, 825.
17. Samuele Romanin, *Storia documentata di Venezia* (Venice, 1853–61), vi. 68–197.
18. Francesco Zanotto, *Storia della Repubblica di Venezia* (Venice, 1864), i. p. vi.
19. Francesco Zanotto, *I Pozzi ed i piombi* (Venice, 1876), 6, 18, 23–4, 44.
20. Camillo Manfroni, 'Gli Studi storici in Venezia dal Romanin ad oggi', *Nuovo Archivio Veneto*, 16 (Venice, 1908), 352–72; 'Bartolomeo Cecchetti': *Dizionario biografico Italiano*; Rinaldo Fulin, *Studi nell archivio degli Inquisitori di Stato* (Venice, 1868), pp. v, 70, 71, and 'Studi sugli Inquisitori di Stato', *Archivio Veneto*, i (Venice, 1871).
21. P. G. Molmenti, *La Storia di Venezia nella vita privata* (Venice, 1880). The 4th edn. was translated into English by Horatio Brown under the title *Venice: Its Individual Growth from the Earliest Beginnings until the Fall of the Republic* (1906–8).
22. J. S. Mill, *Considerations on Representative Government*, ch. 6; Lord Acton, *The History of Freedom and Other Essays* (1907), 49, 213; [Elizabeth Eastlake], 'Venice Defended', *Edinburgh Review*, 146 (July, 1877); [Elizabeth Eastlake], 'Venice: Her Institutions and Private Life', *Quarterly Review*, 168 (Jan. 1889); Horatio F. Brown, *Venice: A Historical Sketch of the Republic*, 2nd edn. (1895), 116, 130, 180, 258; W. Carew Hazlitt, *The Venetian Republic*, 4th edn. (1915), i. p. xv; ii. 568, 615.
23. William Roscoe Thayer, *A Short History of Venice* (New York, 1905), pp. viii, x, 104–7, 109, 116; Francis Marion Crawford, *Gleanings from Venetian History*, 2nd edn. (1907), 111, 142; Mildred Howells, *The Life in Letters of William Dean Howells* (New York, 1928), ii. 122–3.
24. James Grubb, 'When Myths Lose Power: Four Decades of Venetian Historiography', *Journal of Modern History*, 58 (Chicago, 1986), 43–94.
25. Honoré de Balzac, 'Massimila Doni', *Études philosophiques* (Paris, 1845), 2, 10; Galibert, *Venise*, 165; W. H. D. Adams, *The Queen of the Adriatic* (1869), 15; *Hansard*, 204, col. 94; Thayer, *Short History*, p. xi; Byron, *Childe Harold's Pilgrimage*, canto 4, stanza 17; Lindsay Stainton, *Turner's Venice* (1985), 71.
26. *Hansard*, 204, col. 94.
27. Margaret Oliphant, *The Makers of Venice* (1887), 111.
28. Yriarte, *La Vie d'un patricien*, 299; Molmenti, *Storia di Venezia*, 54.
29. J. Ruskin, *The Library Edition of the Works of John Ruskin*, ed. E. T. Cook and A. Wedderburn (1903–12), xxiv. 240.
30. *Edinburgh Review*, 146 (July 1877), 198.
31. Brown, *Historical Sketch*, 'Preface', 260, 310.
32. Thayer, *Short History*, p. xi.
33. Crawford, *Gleanings*, 144–5, 367, 437.
34. *Hansard*, 204, cols. 82, 93.
35. *The Times*, 14 Feb. 1871.
36. Sismondi, *Républiques italiennes*, viii. 130.

37. Daru, *Histoire*, ii. 324; iii. 25–6.
38. Althea Wiel, *The Navy of Venice* (1910), pp. vii, 2, 212, 214, 226, 330.
39. Seeley, *Expansion of England*, 354.
40. Daru, *Histoire*, iii. 209; vi. 282–3.
41. Lamansky, *Secrets d'état*, pp. xiii, 673.
42. Hazlitt, *The Venetian Republic*, ii. 442–3.
43. Carlo Gozzi, *The Memoirs of Count Carlo Gozzi*, trans. J. A. Symonds (1890), pp. x, 4. See also Havelock Ellis, *Affirmations*, 2nd edn. (New York, 1915), 86.
44. *Quarterly Review*, 168 (1889), 71–102.
45. Arnold White, *Efficiency and Empire* (1901), 121.
46. Maurice Barrès, *Un Homme libre*, new edn. (Paris, 1922), 187–91; Symonds, *In the Key of Blue*, 43–52; Carew Martin, 'The Last of the Venetian Masters', *The Art Journal* (1899), 334–8; Pompeo Molmenti, *Tiepolo: la vie et l'œuvre du peintre*, trad. français (Paris, 1911); Heinrich Modern, *Giovanni Battista Tiepolo* (Vienna, 1902); Henry de Chennevières, *Les Tiepolo* (Paris, 1898); Francis Haskell, *Rediscoveries in Art* (1976), 74, 80.
47. For Favretto, see Pavanello and Romanelli (eds.), *Venezia nell'ottocento*. For Hofmannsthal, see Geneviève Bianquis, 'L'Image de Venise dans l'œuvre de Hofmannsthal', *Revue de Littérature comparée*, 32 (Paris, 1958), 321–6.
48. Vernon Lee, *Studies of the Eighteenth Century in Italy*, 2nd edn. (1907), 383.
49. Philippe Monier, *Venice in the Eighteenth Century*, Eng. trans. (1910).
50. Frédéric de Hohenlohe-Waldenburg, *Notes vénitiennes* (Paris, 1899), 130.
51. See Edward Bristow, *Vice and Vigilance* (Dublin, 1977).
52. Virginia Woolf, *The Essays of Virginia Woolf*, ed. A. McNeillie (1986), i. 246.

*Chapter VI*

1. Arthur Symons, *Cities of Italy* (1907), 102; Byron, *Childe Harold's Pilgrimage*, canto 4; Browning, *A Toccata of Galuppi's*; Lee, *Eighteenth Century in Italy*, 101; Charles Dickens, 'An Italian Dream', in *Pictures from Italy*; Hutton, *Venice and Venetia*, 39–40.
2. J. A. Symonds, *Sketches and Studies in Italy and Greece*, new edn. (1905), 281.
3. Flagg, *The City of the Sea*, 53 n.; Burgess, *The Climate of Italy*, 122; *The Times*, 12 Mar. 1910; 7 Oct. 1902.
4. John Chetwode Eustace, *A Classical Tour Through Italy*, 7th edn. (1841), i. 104 n.; *The Times*, 16 Oct. 1875. See also *Builder*, 15 Mar. 1879; *Architect*, 5 June 1880.
5. Benjamin Webb, *Sketches of Continental Ecclesiology* (1848), 299; Henry Christmas, *Shores and Islands of the Mediterranean* (1851), ii. 155–6.
6. Joseph Woods, *Letters of an Architect from France, Italy, and Greece* (1828), i. 256–62; Samuel Rogers, *The Italian Journal of Samuel Rogers*, ed. J. R. Hale (1956), 172–7; Henry Gally Knight, *The Ecclesiastical Architecture of Italy* (1843), i. pl. xxxi; John Edward Bowden, *The Life and Letters of Frederick William Faber* (1869), 102–3; W. E. Gladstone, *Diaries*, ed. M. R. D. Foot (Oxford, 1968– ), i. 533; R. E. Prothero, *The Life and Correspondence of Arthur Penrhyn Stanley*, 4th edn. (1894), i. 265; Eustace, *Classical Tour*, i. 110; Samuel Rogers, *Italy: A Poem* (1830), 48; Smedley, *Sketches from Venetian*

*History*, ii. 293; R. M. Milnes, 'Venice', in *Memorials of a Residence on the Continent* (1838); Gautier, *Voyage*, 17, 55.

7. B. Disraeli, *Contarini Fleming*, new edn. (1888), 207; William Beckford, *Vathek and European Travels*, new edn. (1891), 265.
8. Beckford, *Vathek*, 265; Milnes, 'Venice'; Gautier, *Voyage*, 18; James Fergusson, *The Illustrated Handbook of Architecture* (1855), i. 798; Alphonse de Lamartine, *Voyage en Orient*, new edn. (Paris, 1913), i. 357; ii. 64, 149; Gérard de Nerval, *Œuvres Complètes* (Paris, 1984), ii. 343–4, 396; B. Disraeli, *Tancred*, new edn. (1887), 112, 378.
9. Woods, *Letters*, i. 272, 274; Eustace, *Classical Tour*, i. 115–16; Alfred Barry, *The Life and Letters of Sir Charles Barry* (1867), 51.
10. Henry Russell Hitchcock, *Early Victorian Architecture in Britain* (1954); and 'Victorian Monuments of Commerce', *Architectural Review*, Feb. 1949; Cecil Stewart, *The Stones of Manchester* (1956); John H. G. Archer (ed.), *Art and Architecture in Victorian Manchester* (Manchester, 1985); Barry, *Sir Charles Barry*.
11. Robert Willis, *Remarks on the Architecture of the Middle Ages Especially in Italy* (1835), 1–3; Fergusson, *Illustrated Handbook*, i. pp. lvi–lvii; ii. 766, 796–9; G. E. Street, *Brick and Marble in the Middle Ages* (1855), 166, 258.
12. *The Seven Lamps of Architecture*, ch. 3, sect. 9.
13. *The Stones of Venice*, i. ch. 15; *Seven Lamps*, ch. 2, sect. xi.
14. *The Stones of Venice*, ii. ch. 6.
15. *The Stones of Venice*, ii. ch. 7, sect. 2.
16. *Architect*, 22 Feb. 1879; *Hansard*, 146 (1869), col. 549.
17. *Edinburgh Review*, Jan. 1888.
18. Street, *Brick and Marble*, 165; Charles Eastlake, *A History of the Gothic Revival* (1872), 307; *Building News*, 21 Sept. 1877 (quoted in Michael Brooks, *John Ruskin and Victorian Architecture* (New Brunswick, 1987), 192–3).
19. Eastlake, *Gothic Revival*, 395 ff.; *Building News*, 19 Apr. 1878.
20. John Summerson, *Victorian Architecture: Four Studies in Evaluation* (New York, 1970), 42.
21. Gilbert Scott, *Remarks on Domestic and Secular Architecture Past and Future* (1858), 190–1, 281.
22. *Building News*, 21 Mar. 1879. The reference is to Tower Chambers, Moorgate Street, City of London.
23. Alvise Zorzi, *Venezia scomparsa* (Milan, 1984), 9, 101–10, 120, 130–5; and *Venezia austriaca* (Rome, 1986), 29–32, 50–1; Charles Yriarte, 'Les restaurations de Saint-Marc à Venise', *Revue des Deux Mondes* (Paris), Mar.–Apr. 1880, 827–56.
24. *Illustrated London News*, 29 Jan. 1853; Fergusson, *Illustrated Handbook*, i. p. lvi.
25. Zorzi, *Venezia austriaca*, 120–1, 261; B. Disraeli, *Venetia*, new edn. (1881), 372–3; Marilyn Perry, 'Antonio Sanquirico, Art Merchant of Venice': *Labyrintos*, 1–2 (Florence, 1983), 67–111; National Trust, *Kingston Lacy, Dorset* (1987), 38, 45; Viola Bankes, *A Dorset Heritage* (1953), 178; *Illustrated London News*, 22 Oct. 1892; Augustus Hare and St Clair Baddeley, *Venice*, 7th

edn. (1907), 11; Elizabeth Eastlake, *Journals and Correspondence*, ed. C. E. Smith (1895), ii. 19–20; Achille Bosisio, 'La fuga di un capolavoro di Paolo Veronese in Inghilterra': *Giornale economico della camera di commercio di Venezia* (Venice), Jan.–Feb. 1970, 27–31; Hazlitt, *The Venetian Republic*, i. 361; Mario Dalla Costa, *La Basilica di San Marco e i restauri dell 'ottocento* (Venice, 1983), 64; Journal of Enid Layard: BL Add. MSS 46162, fo. 125.

26. See Roland Mortier, *La Poétique des ruines en France* (Geneva, 1974).
27. F. R. de Chateaubriand, *Mémoires d'outre tombe*, new edn. (Paris, 1964), ii. 772.
28. E. Viollet-le-Duc, *Dictionnaire raisonné de l'architecture française du* XI^e^ *au* XVI^e^ *siècle* (Paris, 1866), viii. 14–34; Nikolaus Pevsner, 'Scrape and Anti-Scrape', in Jane Fawcett (ed.), *The Future of the Past* (1976).
29. Kent Robert Greenfield, 'Commerce and Enterprise at Venice, 1830–1848': *Journal of Modern History*, 11 (Chicago, 1939), 313–33.
30. Zorzi, *Venezia scomparsa*, 121, 140; C. Robotti, 'Le idee di Ruskin ed i restauri della Basilica di San Marco', *Bolletino d'Arte*, 1–2 (Rome, 1976), 115–21; Otto Demus, *The Church of San Marco in Venice* (Washington, 1960), 102, 196–7; Thomas Okey, *Old Venetian Palaces and Old Venetian Folk* (1907), 24; Dalla Costa, *La Basilica di San Marco*, 12–26, 36–8.
31. Vincenzo Fontana, 'Camillo Boito e il restauro a Venezia', *Casabella*, 472 (Milan, 1981); Dalla Costa, *La Basilica di San Marco*, 111–12. For Boito see *Dizionario Biografico Italiano* and Carroll Meeks, *Italian Architecture 1750–1914* (New Haven, Conn., 1966).
32. [Pietro Saccardo], *San Marco, gl'Inglesi e noi* (Venice, 1879), 6–8; Robotti, 'Le idee di Ruskin'; Dalla Costa, *La Basilica di San Marco*, 36; E. Viollet-le-Duc, 'De la restauration des anciens édifices en Italie', *Encyclopédie de l'architecture*, 2nd ser., i (Paris, 1872); *Building News*, 14 Dec. 1877; Mary Shelley, *Rambles in Germany and Italy* (1844), ii. 79; George Stillman Hillard, *Six Months in Italy* (1853), i. 30; B. Disraeli, *Contarini Fleming*, new edn. (1888), 206; Gautier, *Voyage en Italie*, 80; Mark Twain, *The Innocents Abroad* (New York, 1869), i. ch. 22.
33. John Ruskin, *Ruskin in Italy: Letters to his Parents 1845*, ed. H. J. Shapiro (Oxford, 1972); *The Stones of Venice*, ii. ch. 6, sect. 11; iii. ch. 4, sect. 1; ii. ch. 7, sect. 1; iii. ch. 2, sect. 90.
34. *The Stones of Venice*, ii. ch. 7, sect. 47; iii. ch. 4, sect. 36; Ruskin, *The Library Edition*, ix. 11; x. 459; xxii. 523–5.
35. There is a possibility that *Seven Lamps* influenced the first main example of the genre, Butterfield's All Saints' Westminster. See Hitchcock, *Early Victorian Architecture*, 572–86; Paul Thompson, *William Butterfield* (1971), 163–5.
36. James Fergusson, *History of the Modern Styles of Architecture*, 2nd edn. (1873, repr. as 3rd edn. 1891), 43–9; 102–15; John T. Emmett, *Six Essays*, repr. (1972); Summerson, *Victorian Architecture*, 2–10; Brooks, *John Ruskin and Victorian Architecture*, 209; T. G. Jackson, *Modern Gothic Architecture* (1873), 3; and *The Recollections of Thomas Graham Jackson*, ed. Basil Jackson (1950), 119; Andrew Saint, *Richard Norman Shaw* (New Haven, Conn., 1976); W. R.

Lethaby, *Philip Webb and his Work* (1935); Mark Swenarton, *Artisans and Architects* (1989), 52–6; Kenneth Clark, *The Gothic Revival*, new edn. (1974), 7.

37. Lethaby, *Philip Webb*, 144; Brooks, *John Ruskin and Victorian Architecture*, 272; *Athenaeum*, 23 June 1877; Charles Dellheim, *The Face of the Past: The Preservation of the Medieval Inheritance in Victorian England* (Cambridge, 1982), 91.
38. Marcel Proust, *La Correspondance de Marcel Proust*, ed. Philip Kolb (Paris, 1976–93), ii. 385; iii. 180; vii. 237; xi. 273; John Ruskin, *La Bible d'Amiens*, trans. Marcel Proust, repr. (Paris, 1947), 91; Jean Autret, *Ruskin and the French before Marcel Proust* (Geneva, 1965), 13–18, 23–37, 130–1.
39. Robotti, 'Le idee di Ruskin'; Zorzi, *Venezia scomparsa*, 140; Dalla Costa, *La Basilica di San Marco*, 43–108; Count A. Zorzi, 'Ruskin in Venice, I', *Cornhill Magazine*, 102 (Aug. 1906), 250–65, and 'Ruskin in Venice, II', *Cornhill Magazine*, NS 21 (Nov. 1906), 366–79. See also Jeanne Clegg, 'John Ruskin's Correspondence with Angelo Alessandri', *Bulletin of the John Rylands University Library*, 60 (1977–8), 404–33.
40. *L'Avvenire dei monumenti a Venezia* (Venice, 1882), 6–9, 11, 18. W. D. Caroë, who knew Boni well, said that he had drafted this pamphlet. See *Builder*, 19 July 1884. Boni also published a signed article, 'L'Avvenire dei nostri monumenti', *Gazzetta di Venezia*, 8 Dec. 1882. On Boni generally see Eva Tea, *Giacomo Boni nella vita del suo tempo*, 2 vols. (Milan, 1932); Eva Tea, 'Corrispondenza fra Philip Webb e Giacomo Boni', *Annales Institutorum*, 13 (Rome, 1940–1), 127–48; 14 (Rome, 1942–3), 135–205; and 'Il Carteggio Boni–Caroë sui monumenti Veneziani', *Archivi*, 26 (Rome, 1959), 234–54.
41. Lionello Puppi, 'Qualche materiale, e una reflessione, nel restauro architettonico secondo Camillo Boito', *Antichita Viva*, 21 (Florence, 1982), 75–9; Fontana, 'Camillo Boito e il restauro a Venezia'.
42. 'Je hais le mouvement qui déplace les lignes' (Baudelaire, 'La Beauté', in *Les Fleurs du Mal*). Proust, among others, perceived that the leading Romantics were Classicists at heart. See John Ruskin, *Sésame et les lys*, trans., notes, and preface by Marcel Proust, 10th edn. (Paris, 1935), 51–2.
43. *The Stones of Venice*, ii. ch. 6, sect. 11. For 18th-c. aesthetics see Samuel H. Monk, *The Sublime: A Study of Critical Theories in Eighteenth-Century England* (New York, 1935), and Walter John Hipple, *The Beautiful, The Sublime and the Picturesque in Eighteenth-Century Aesthetic Theory* (Carbondale, Ill., 1957).
44. Maurice Barrès, 'Amori et Dolori Sacrum': *L'Œuvre de Maurice Barrès* (Paris, 1967), vii. 16.
45. Proust, *Correspondance*, viii. 288; *L'Avvenire dei monumenti a Venezia*, 7; Giacomo Boni, 'Venezia imbellata', cited in Zorzi, *Venezia scomparsa*, 140.
46. The first expression was coined by Jules Barbey-d'Aurevilly; the second by J. K. Huysmans. Both writers are discussed in Mario Praz, *The Romantic Agony*, 2nd edn. (1970).
47. Jean Pierrot, *L'Imaginaire décadent, 1880–1900* (Paris, 1977), 20.
48. Ruskin, *La Bible d'Amiens*, trans. Marcel Proust, 84. See also pp. 64, 78–80.

*Chapter VII*

1. Arthur Symons, 'The Music of Venice', The *Saturday Review*, 17 Oct. 1908; C. E. Norton, *Historical Studies of Church-Building in the Middle Ages* (New York, 1880), 41. See also Raymond Schwab, *The Oriental Renaissance* (New York, 1984); Edward Said, *Orientalism* (Harmondsworth, 1985).
2. Erwin Panofsky, '*Et in Arcadia Ego*: On the Conception of Transience in Poussin and Watteau', in R. Kiblansky and H. J. Paton (eds.), *Philosophy and History: Essays Presented to Ernst Cassirer* (Oxford, 1936).
3. A. W. Buckland, *The World Beyond the Esterelles* (1884), ii. 209; H. Brown, *Life on the Lagoons*, 5th edn. (1909), 73.
4. Brown, *Life on the Lagoons*, 207; Marcel Proust, *Textes Retrouvés*, ed. Philip Kolb and Larkin Price (Urbana, Ill., 1968), 167; Guillaume Apollinaire, *Œuvres en prose complètes*, ii (Paris, 1991), 720; Hohenlohe-Waldenburg, *Notes vénitiennes*, 5 (to facilitate quotation this sentence has been reversed).
5. J. A. Symonds, *Sketches and Studies in Italy and Greece*, 1st ser., new edn. (1905), 255.
6. See *Builder*, 13 Apr. 1889.
7. *Athenaeum*, 15 Jan. 1876; William Morris, *The Collected Letters of William Morris*, ed. Norman Kelvin (Princeton, NJ, 1984), i. 541 n.; *The Times*, 13 Nov. 1879, 22 Sept. 1885; Zorzi, *Venezia scomparsa*, 135; Robert Browning, *Learned Lady*, 168; *Builder*, 13 Apr. 1889.
8. *Builder*, 28 July 1883, 26 Sept. 1885; *Athenaeum*, 2 July 1881; *The Times*, 3 and 18 Aug. 1883, 5 Oct. 1886; Crawford, *Gleanings from Venetian History*, 7.
9. *The Times*, 18 Nov. 1879.
10. W. J. Stillman, *The Autobiography of a Journalist* (1901), ii. 238; Dalla Costa, *La Basilica di San Marco*, 57; *Builder*, 16 and 30 Jan. 1869; William Morris, *Letters*, i. 541 n.; *Building News*, 14 Dec. 1877; Yriarte, 'Les Restaurations de Saint-Marc à Venise'; Horatio F. Brown, 'Recent Restorations of the Basilica', *The Times*, 23 Feb. 1914.
11. G. E. Street, Report in *Architect*, 22 May 1880; [Saccardo], *San Marco, gl'Inglesi e noi*; Anon., *La Basilica di San Marco in Venezia nel suo passato e nel suo avvenire* (Venice, 1883), 17.
12. Tea (ed.), 'Il Carteggio Boni-Caroë'; *Building News*, 5 Dec. 1879, 13 and 27 Feb. 1880; Della Costa, *La Basilica di San Marco*, 32–3; *The Times*, 22 and 27 Nov. 1879; SPAB, *Report of the Third Annual Meeting*, June 1880.
13. *The Times*, 22 and 28 Nov. 1879 (letter from E. Poynter, quoting *Il Rinnovamento*); William Morris, *Collected Letters*, i. 486–7, 541, 576, 588–9; *Athenaeum*, 10 Nov. 1877, 20 Nov. 1880; Papers of the St Mark's Committee, BL Add. MSS 38831; Camillo Boito, 'I restauri di San Marco', *Nuova Antologia di Scienze, Lettere ed Arti*, 18, ser. 2 (Rome, Dec. 1879), 701–21.
14. *The Times*, 17, 18, and 19 Nov. 1879; *Building News*, 28 June 1878, 14 Nov. 1879, 17 Dec. 1880; *Builder*, 15 Nov. 1879; *Architect*, 3 Jan. and 7 Feb. 1880.
15. *The Times*, 8, 21, and 22 Nov. 1879; *Building News*, 21 Nov. 1879; Yriarte, 'Les Restaurations de Saint-Marc à Venise'; Norton, *Church-Building in the Middle Ages*, 57–8.

16. *The Times*, 5 Aug., 21 Sept., and 19 Oct. 1886; *Builder*, 4 Dec. 1886; Anon., *La Basilica di San Marco in Venezia*, 11; SPAB, *Report of the Eleventh Annual Meeting*, July 1888; Tea (ed.), 'Il Carteggio Boni-Caroë'; Tea (ed.), 'Corrispondenza fra Philip Webb e Giacomo Boni (I & II)'.
17. *Builder*, 29 Mar. 1902; H. F. Brown's report on Marangoni's operations, *The Times*, 23 Feb. 1914; F. Forlati, 'The Work of Restoration in San Marco', in Otto Demus, *The Church of San Marco in Venice* (Washington, DC, 1960).
18. John Ruskin, *Ruskin's Letters from Venice, 1851–1855*, ed. John Lewis Bradley (New Haven, Conn., 1955), 128.
19. Boni's report in *Builder*, 28 June 1879; Stillman's reports in *The Times*, 12 Aug. 1886, 16 Sept. 1889; Brown, *Life on the Lagoons*, 185–9.
20. Tea (ed.), 'Corrispondenza fra Philip Webb e Giacomo Boni (II)'. Saccardo's paper is summarized by Thomas Okey, *The Old Venetian Palaces and Old Venetian Folk* (1907), 33–8.
21. *Builder*, 1 Aug. 1885; *The Times*, 16 Sept. 1889; Brown, *Life on the Lagoons*, 186; Thomas Okey, *Venice and its Story* (1905), 317; Gregory, *Seventy Years*, 172; Journal of Enid Layard, BL Add. MSS 46163, fo. 18.
22. *Building News*, 21 Nov. 1879; *Architect*, 7 Feb. and 19 June 1880; Flagg, *The City of the Sea*, 49; Horatio F. Brown, *In and Around Venice* (1905), 59; Boito, *Storielle vane*, 183.
23. *The Times*, 15 July 1902; *Builder*, 19 July 1902; *Architect and Contract Reporter*, 29 Aug. 1902.
24. *Builder*, 26 July and 30 Aug. 1902; Laura Ragg, *Crises in Venetian History* (1928), 257–67; Consiglio Communale di Venezia, *Il Campanile di San Marco riedificato* (Venice, 1912); *The Times*, 27 Apr. 1903; Viola Meynell (ed.), *Friends of a Lifetime: Letters to Sydney Carlyle Cockerell* (1940), 163; Roger Fry, *The Letters of Roger Fry*, ed. Denys Sutton (1972), i. 196; Letters of Henry Wallis, Bodleian MSS don. e. 79, fo. 193; Ragg, *Things Seen in Venice*, 92; Hugh A. Douglas, *Venice on Foot* (1907), 18; Hutton, *Venice and Venetia*, 89–90.
25. *The Times*, 26, 27, and 29 April 1912; *Builder*, 30 Aug. 1912.
26. *Builder*, 30 Aug. 1902.
27. E. V. Lucas, *A Wanderer in Venice*, 8th edn. (1925), 41–2, 43; *Architect and Contract Reporter*, 1 Aug. 1902; Maurice Barrès, 'Dix jours en Italie', in *L'Œuvre de Maurice Barrès* (Paris, 1967), ix. 69–121.
28. Cecil Torr, *Small Talk at Wreyland*, 3rd ser. (1923), 53.
29. F. T. Marinetti, *Le Futurisme*, new edn. (Lausanne, 1980), 94.
30. F. T. Marinetti *et al.*, *Marinetti e il futurismo*, ed. Luciano de Maria (Verona, 1973), 125, 141.
31. Marinetti, *Marinetti e il futurismo*, 27, 29.
32. Marinetti, *Le Futurisme*, 93, 94; Marinetti, *Marinetti e il futurismo*, 84, 220.
33. Marinetti, *Marinetti e il futurismo*, 29.
34. The Avant-Garde and its relation to Futurism are discussed in Theda Shapiro, *Painters and Politics: The European Avant-Garde and Society, 1900–1925* (New York, 1976); Patricia Leighton, *Reordering the Universe: Picasso and Anarchism, 1897–1914* (Princeton, NJ, 1989); Kenneth Silver, *Esprit de Corps: The Art of the Parisian Avant-Garde and the First World War* (Princeton, NJ, 1989); Mark Antliff, *Inventing Bergson: Cultural Politics and the Parisian*

*Avant-Garde* (Princeton, NJ, 1993); Ivor Davis, 'Western European Art Forms Influenced by Nietzsche and Bergson', *Art International*, 29:3 (Lugano, March 1975), 49–55; Ester Coen, 'The Violent Urge Towards Modernity: Futurism and the International Avant-Garde', in Emily Braun (ed.), *Italian Art in the Twentieth Century* (Munich, 1989). For Marinetti's political career see James Joll, *Three Intellectuals in Politics* (New York, 1960).

35. Marinetti, *Marinetti e il Futurismo*, 28–9, 162–3; Marinetti, *Le Futurisme*, 95, 121.
36. Hutton, *Venice and Venetia*, 39. For the Bloomsbury view see Clive Bell, *Art*, new edn. (1949), and Roger Fry, *Vision and Design* (1923). The quotation is Forster's. Gide's comment is in his *Journal, 1889–1939* (Paris, 1951), 152. For Kandinsky, see Coen, 'The Violent Urge Towards Modernity'.
37. Rosebery MSS, NLS 10127, fo. 4.
38. Damerini, *D'Annunzio e Venezia*, 78–9, 266–8.
39. Quotations from *Trionfo della morte* and *Il Fuoco*. See *Prose di romanzi*, i (Milan, 1988), 673; ii (Milan, 1989), 233, 253. For a discussion in English of d'Annunzio and the politics of nostalgia see Richard Drake, *Byzantium for Rome* (Chapel Hill, NC, 1980).

*Chapter VIII*

1. D'Annunzio, *Prose di romanzi*, ii. 245; Walter Pater, *The Renaissance: Studies in Art and Poetry*, new edn. (1888), 156–7; Bernhard Berenson, *The Italian Painters of the Renaissance*, new edn. (1960), 30; Proust, *À la recherche du temps perdu*, iii. 626.
2. H. F. Brown, *John Addington Symonds: A Biography* (1895), i. 326–8; Symonds, *Letters*, i. 424; J. A. Symonds, *Renaissance in Italy*, iii (2nd edn., 1882), 370, 380; Ruskin, *The Library Edition*, xxii. 77; A. H. Layard (ed.), *The Italian Schools of Painting Based on the Handbook of Kugler* (4th edn., 1907), ii. 612 n.; Berenson, *Italian Painters of the Renaissance*, 9.
3. William Morris, *Selected Writings and Designs*, ed. Asa Briggs (1962), 32. For the intellectual background in Germany and Austria see Fritz Stern, *The Politics of Cultural Despair* (Berkeley, Calif., 1961), and William McGrath, *Dionysian Art and Populist Politics* (New Haven, Conn., 1974).
4. Paul Bourget, *Essais de psychologie contemporaine*, new edn. (Paris, 1920), pp. xxv–xxvi.
5. Jacob Burckhardt, *The Letters of Jacob Burckhardt*, trans. Alexander Dru (New York, 1955), 205.
6. Anne Thackeray, *Miss Angel* (1875), 45; Julia Cartwright, 'The Artist in Venice', *The Portfolio* (1883–4), 37–42; Ouida, *Santa Barbara and Other Tales* (1891), 12–13; [Elizabeth Eastlake], 'Venice Defended', *Edinburgh Review*, July 1877; Brown, *Life on the Lagoons*, 291–2; Stopford Brooke, *The Sea Charm of Venice* (1907), 90–1; Proust, *À la recherche du temps perdu*, iii. 640.
7. Yriarte, *La vie d'un patricien*, 16; Symonds, *Renaissance in Italy*, iii. 347–8; Zorzi, *Venezia scomparsa*, 147; *Builder*, 13 Apr. 1889.
8. Ruskin, *Modern Painters*, v. ch. 9, sect. 1.
9. Morris, *Selected Writings*, 85, 87.

10. May Morris, *William Morris: Artist, Writer, Socialist* (Oxford, 1936), i. 145.
11. Burckhardt, *Letters*, 97.
12. W. R. Lethaby, *Philip Webb and his Work* (1935), 185; E. P. Thompson, *William Morris, Romantic to Revolutionary*, new edn. (1977), 663, 805; Thomas Okey, *A Basketful of Memories* (1930), 123–4; Shapiro, *Painters and Politics*, 126.
13. James Buzard, 'The Uses of Romanticism: Byron and the Victorian Continental Tour', *Victorian Studies* (Indianapolis), autumn 1991.
14. Matthew Arnold, *Essays in Criticism*, 1st ser., new edn. (1928), 192; Chateaubriand, *Mémoires d'outre tombe*, ii. 1020 n.
15. Symonds, *Letters and Papers*, 223; Browning, *Letters Collected by T. J. Wise*, 306–7; Okey, *A Basketful of Memories*, 70–2, 83, 117–25; Symonds, *Letters*, iii. 680–1; S. A. Barnett, *Canon Barnett: His Life, Work, and Friends* (1918), 361–3.
16. Marie and Squire Bancroft, *Recollections of Sixty Years* (1909), 201–6; Dudley Harbron, *The Conscious Stone* (1949), 104; Ellen Terry, *The Story of My Life* (1908), 106; Joseph Knight, *Recollections* (1893), 26–7, 303–5; Laurence Irving, *Henry Irving: The Actor and his World* (1951), 333; Hesketh Pearson, *Beerbohm Tree: His Life and Laughter* (1956), 118.
17. *Builder*, 2 Jan. 1892; *The Times*, 7 June 1892; Symonds, *Letters*, iii. 689.
18. *Builder*, 3 Sept. 1887; Margaret Oliphant, *The Makers of Venice* (1887), 389; *The Times*, 24 Aug. 1887.
19. BL Add. MSS 52748, fo. 282.
20. Zorzi, *Venezia scomparsa*, 145–7; Damerini, *D'Annunzio e Venezia*, 266–7; *Builder*, 13 Apr. 1889; Tea (ed.), 'Corrispondenza fra Philip Webb e Giacomo Boni (II)'.
21. James, *The House of Fiction*, 41, 147–9.
22. BUL DM/376.
23. Horatio Brown, 'Venice in Wartime', *Cornhill Magazine*, Mar. 1917; Horatio Brown, 'Venice in Wartime', *The Times*, 25 Jan. 1917; Ragg, *Crises in Venetian History*, 270–86; Reports in *The Times*, 2 Nov. 1915, 20 Sept. 1916; Damerini, *D'Annunzio e Venezia*, 142–3; Barrès, 'Dix jours en Italie'; Colette, 'Un Taube sur Venise', *Œuvres* (Paris, 1986), ii. 536–8.
24. Henri de Régnier, 'Venise Menacée', *Revue des Deux Mondes*, 42 (Nov.–Dec. 1927), 88–93 (subsequently repr. in *L'Altana*).
25. Massimo Cacciari, 'Venezia postuma', in Pavanello and Romanelli, *Venezia nell'ottocento*.
26. Brown, 'Venice in Wartime', *Cornhill Magazine*.
27. *The Times*, 8 Jan. 1918.
28. Apollinaire, *Œuvres en prose complètes*, ii. 12–13.

*Epilogue*

1. *Revue des Deux Mondes*, 42 (Paris, 1927), 403.
2. Cecil Roberts, *The Bright Twenties* (1970), 140–4.
3. *The Times*, 2 Nov. 1915.
4. Ibid., 3 Sept. 1918.
5. NLS MS 10126, fos. 82, 232.

6. Ojetti, *Cose viste*, i. 463.
7. *Revue des Deux Mondes*, 42 (Paris, 1927), 401.
8. Cyril Connolly, *Enemies of Promise* (1938), 278.
9. Ernest Rhys, *Everyman Remembers* (1931), 308.
10. Preface to *Some Imagist Poets* (1915).
11. Connolly, *Enemies*, 93–4, 107.
12. Quoted in Williams, *Culture and Society*, 222.
13. Colette, 'Un entretien avec un Prince de Hohenlohe', *Œuvres*, ii. 543–4; Ojetti, *Cose viste*, i. 238.
14. NLS MS 10126, fo. 82.
15. Ragg, *Crises in Venetian History*, 285–6; *The Times*, 11 June 1938, 12 Jan. 1962; Luhan, *European Experiences*, 290–300; Ragg, 'Venice when the Century Began'; Henry James, *Letters*, iv. 753.

# INDEX

*Note*: Churches are listed together under that entry